163
Advances in Biochemical Engineering/Biotechnology

Aims and Scope

This book series reviews current trends in modern biotechnology and biochemical engineering. Its aim is to cover all aspects of these interdisciplinary disciplines, where knowledge, methods and expertise are required from chemistry, biochemistry, microbiology, molecular biology, chemical engineering and computer science.

Volumes are organized topically and provide a comprehensive discussion of developments in the field over the past 3–5 years. The series also discusses new discoveries and applications. Special volumes are dedicated to selected topics which focus on new biotechnological products and new processes for their synthesis and purification.

In general, volumes are edited by well-known guest editors. The series editor and publisher will, however, always be pleased to receive suggestions and supplementary information. Manuscripts are accepted in English.

In references, Advances in Biochemical Engineering/Biotechnology is abbreviated as *Adv. Biochem. Engin./Biotechnol.* and cited as a journal.

More information about this series at http://www.springer.com/series/10

Ulrich Martin • Robert Zweigerdt • Ina Gruh
Editors

Engineering and Application of Pluripotent Stem Cells

With contributions by

P. W. Andrews · M. J. Avaca · M. Burcin · X. Y. Chan ·
K. Christensen · R. David · M. B. Elliott · S. Gerecht ·
I. Gruh · J. A. Halliwell · F. Hausburg · J. Itskovitz-Eldor ·
J. J. Jung · H. Kempf · O. Kyriakides · N. Lavon · B. Macklin ·
U. Martin · S. Merkert · C. Patsch · F. Roudnicky ·
T. M. Schlaeger · M. Zimerman · R. Zweigerdt

 Springer

Editors
Ulrich Martin
Hannover Medical School (MHH)
Leibniz Research Labs for Biotechnology
Hannover, Germany

Robert Zweigerdt
Hannover Medical School (MHH)
Leibniz Research Labs for Biotechnology
Hannover, Germany

Ina Gruh
Hannover Medical School (MHH)
Leibniz Research Labs for Biotechnology
Hannover, Germany

ISSN 0724-6145   ISSN 1616-8542   (electronic)
Advances in Biochemical Engineering/Biotechnology
ISBN 978-3-319-73590-0   ISBN 978-3-319-73591-7   (eBook)
https://doi.org/10.1007/978-3-319-73591-7

Library of Congress Control Number: 2018932417

Printed on acid-free paper

This Springer imprint is published by Springer Nature
The registered company is Springer International Publishing AG
The registered company address is: Gewerbestrasse 11, 6330 Cham, Switzerland

Preface

We are very pleased to present this volume on "Engineering and Application of Pluripotent Stem Cells" in *Advances in Biochemical Engineering and Biotechnology*. Human pluripotent stem cells (PSCs), including embryonic stem cells (ESCs) and induced pluripotent stem cells (iPSCs), have far-reaching potential for proliferation and differentiation. Thus, their availability offers novel opportunities in basic research, disease modelling, drug discovery, toxicology studies, and the development of cellular therapies. For all of these applications, the engineering of pluripotent stem cells in different forms is required or at least helpful.

Engineering can take place on different levels, starting with the generation of induced PSCs through the targeted molecular reprogramming of somatic cells – an approach that has also stimulated direct cell fate conversion of somatic cells into another differentiated phenotype, thereby bypassing the pluripotent state. Other aspects of PSC engineering are novel approaches in gene editing, treatment with growth factors and small molecules or cell seeding on nanostructures for stepwise differentiation, static or continuous mass culture in various types of bioreactors, and cultures in three-dimensional matrices and their vascularization.

In the clinical applications of cellular products, the regulatory requirements for standardization and safety are demanding. GxP-conforming protocols have to be developed; furthermore, potential risks (e.g., teratoma formation) have to be assessed and minimized. Further research on the genetic stability of PSCs is also required, especially because genetic abnormalities in their therapeutic derivatives might carry a risk of tumor formation.

To prepare this volume, contributions from leading researchers and experts in specific fields of basic and applied stem cell research have been assembled, including investigators from the pharmaceutical industry. The methodological aspects of PSC engineering, along with critical discussions of technical limitations and risks, are the focus of most contributions. These chapters address important current aspects in the field, covering the generation of transgene-free induced pluripotent cells, lineage-specific differentiation and enrichment, as well as the mass production of respective derivatives via large-scale culture protocols. Recent developments in stem cell-based tissue engineering, progress and challenges in their use for drug screening, and discussions of safety aspects round off the volume.

We hope that this collection of reviews will be useful not only for stem cell researchers but particularly for investigators in related fields, including physicians, chemists, and engineers who intend to enter the field of stem cell research. In addition, we anticipate that this volume will provide a basic reading for students who want to deepen their knowledge on the field of PSC biology and regenerative medicine.

Finally, we would like to thank all of authors for their excellent contributions, as well as Springer for implementation of this project. We would like to specifically thank Sandra Stelljes for coordination of the collection of articles and the review process, as well as Prof. Thomas Scheper and especially Ms. Alamelu Damodharan for their patience and excellent work as production editors.

Hannover, Germany   Ulrich Martin
August 2017   Robert Zweigerdt
                                                                             Ina Gruh

Contents

Adv Biochem Eng Biotechnol (2018) 163: 1–22
DOI: 10.1007/10_2017_29
© Springer International Publishing AG 2017
Published online: 27 October 2017

Nonintegrating Human Somatic Cell Reprogramming Methods

Thorsten M. Schlaeger

Abstract Traditional biomedical research and preclinical studies frequently rely on animal models and repeatedly draw on a relatively small set of human cell lines, such as HeLa, HEK293, HepG2, HL60, and PANC1 cells. However, animal models often fail to reproduce important clinical phenotypes and conventional cell lines only represent a small number of cell types or diseases, have very limited ethnic/genetic diversity, and either senesce quickly or carry potentially confounding immortalizing mutations. In recent years, human pluripotent stem cells have attracted a lot of attention, in part because these cells promise more precise modeling of human diseases. Expectations are also high that pluripotent stem cell technologies can deliver cell-based therapeutics for the cure of a wide range of degenerative and other diseases. This review focuses on episomal and Sendai viral reprogramming modalities, which are the most popular methods for generating transgene-free human induced pluripotent stem cells (hiPSCs) from easily accessible cell sources.

T.M. Schlaeger (✉)
Stem Cell Program, Boston Children's Hospital, Karp RB09213, 1 Blackfan Circle, Boston, MA 02446, USA
e-mail: thorsten.schlaeger@childrens.harvard.edu

Graphical Abstract

Nucleofection of Episomal Plasmids

Sendai Viral Transduction

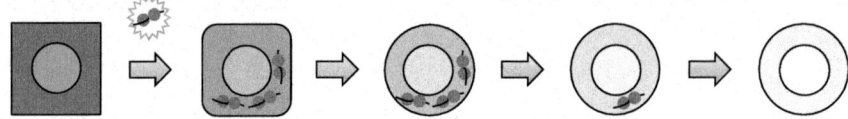

Repeated Lipofection of Modified mRNAs

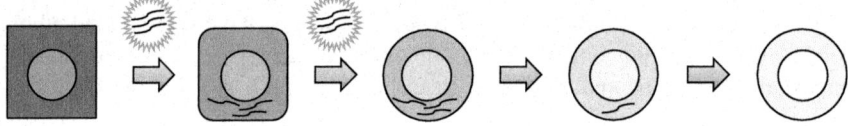

Somatic Cell   *Reprogramming Intermediates*   *iPSC*

Keywords Episomal, hiPSCs, Reprogramming, Sendai virus

Contents

Abbreviations

cGMP   current Good Manufacturing Practice
ECC   Embryonal carcinoma cell
Epi   Episomal
ESC   Embryonic stem cell
hESC   Human embryonic stem cell
hiPSC   Human induced pluripotent stem cell

mESC   Mouse embryonic stem cell
OSKM   Oct4/Sox2/Klf4/c-Myc
PBMC   Peripheral blood mononuclear cell
SeV   Sendai virus/Sendai viral
TAD   Trans-activating domain

1 Introduction: From Embryonal Carcinoma to Induced Pluripotent Stem Cells

The roots of pluripotent stem cell biology extend back to the seventeenth century when descriptions of bizarre tumors began to appear in the early peer-reviewed medical literature [1]. In his seminal textbook on cancer, Virchow (1863) classified these types of malformations as "teratoid" tumors [2], from the Greek word for "monstrous." Indeed, these tumors often have a grotesque appearance as they may contain a variety of tissues types, including hair, skin, muscle, bone, eye, lung, gut, and brain-like structures. However, their nature remained mysterious until modern research tools became available and enabled more detailed studies.

Through pioneering single-cell transplantation experiments, Pierce and Kleinsmith discovered that some teratocarcinomas contain pluripotent cancer stem cells within the undifferentiated (embryonal carcinoma-like) regions of the tumor [3]. When they injected these so-called embryonal carcinoma cells (ECCs) into mice, they found that single ECCs could generate new tumors. Importantly, these tumors again comprised regions containing undifferentiated embryonal carcinoma-like regions as well as regions containing differentiated teratoma-like tissues that often included derivatives of all three germ layers [3]. Martin and Evans showed that ECCs can be coaxed in vitro to differentiate into embryoid bodies that likewise contain a variety of cell types and tissue structures and whose development mimics normal ontogeny [4]. However, ECCs are malignant and frequently have an abnormal karyotype; when injected into blastocyst-stage embryos, ECCs tend to produce abnormally formed embryos that often contain tumors [5, 6]. Nevertheless, some of the cells are able to respond properly to developmental cues provided by the growing embryo and participate in normal tissue formation [7, 8].

Injecting ECCs is not the only method of generating experimental teratomas or teratocarcinomas. In the early 1900s, Askanazy showed that transplanting normal embryonic tissues into ectopic sites results in the formation of pluripotential tumors [9]. Indeed, as early as 1683, Tyson [1] had speculated that teratomas originate from embryo-like material and thus essentially represent normal development gone awry. The notion that normal (i.e., not tumor-derived) pluripotent stem cells probably exist in the early embryo was confirmed by embryonic tissue transplantation studies carried out by Stevens [10] and Solter [11]. A major step toward understanding and harnessing the developmental potential of these cells was taken in 1981 by Evans and Kaufman [12] and Martin [13] when they reported the first successful isolation and culture of normal (i.e., nonmalignant) pluripotent stem cells from early mouse embryos. These cells, derived from the inner cell mass of pre-implantation mouse embryos, were dubbed mouse embryonic stem cells (mESCs) by Martin. When used to create chimeric embryos, mESCs can generate virtually all cell types, including germ cells [14], and even entire mice [15]. These features, combined with the ability to modify genes in mESCs precisely, allowed Smithies [16] and Capecchi [17] to launch the knockout mouse revolution [18].

However, it was almost two decades (and not for a lack of trying [19]) from the development of mESC technologies until the first successful derivation of human embryonic stem cells (hESCs) by Thomson in 1998 [20]. A contributor to this significant delay was the fact that it was then unknown that hESCs represent a different developmental state (primed pluripotent post-implantation epiblast for hESCs versus naive pluripotent pre-implantation inner cell mass/epiblast for mESCs) [21, 22], with different cytokine and cell culture requirements. Early protocols for the derivation of mESCs failed to yield ESCs from many species, and many mouse strains were refractory until the conditions were refined [23]. Other, more persistent challenges were posed by the limited access to human embryos and the research and funding restrictions associated with the production and use of hESCs. Moreover, conventional hESC derivation methods cannot generate cell lines from existing patients. Again, more than a decade passed before the first patient-specific hESCs were generated by somatic cell nuclear transfer [24], a technology that is technically demanding and comes with additional challenges, such the scarcity of human eggs available for such experiments and the ethical concerns, held by some [25], associated with creating and then destroying human embryos for research purposes.

The literature is awash with reports claiming that other types of allegedly pluripotent cells exist as part of normal biology in various fetal or adult tissues, or that they can be obtained from mature tissue by simple culture methods. It is beyond the scope of this review to discuss the true stemness, pluripotency, or utility of the cells described in these reports. The reader is encouraged to refer to recent reviews of the characteristics, markers, and standards of assessment of genuine pluripotent stem cells (e.g. [26–28]). In adults, most somatic cells are differentiated and have very limited proliferative or developmental potential. This fate choice was long believed to be reinforced by irreversible epigenetic mechanisms [29]. The successful cloning of frogs by Gurdon [30] and adult sheep by Wilmut and colleagues [31] therefore came as quite a surprise because they demonstrated that the normal course of somatic cell development and differentiation is reversible, at least when the epigenetic memory of a differentiated cell's nucleus becomes erased and reprogrammed through the action of factors present in oocytes. Pluripotent stem cells likewise express trans-acting factors that can reprogram somatic cells to a pluripotent state, as shown by cell fusion studies in which the resulting somatic–pluripotent cell hybrids quickly reached a stable pluripotent stem cell-like state [32–36]. One important implication of these studies is that the pluripotency-inducing factors and gene regulatory networks can dominate over their somatic cell counterparts that normally act to maintain the differentiated cell state.

About three decades ago, the first trans-acting factors that could alter the fate and differentiation state of somatic cells were discovered. A textbook case in point is the *Drosophila* homeobox gene Antennapedia, which, when expressed in the embryonic domain that normally gives rise to antennae, reprograms the tissue to form ectopic legs instead of antennae [37]. Another early example of this new class of transcription factors (often called master regulator transcription factors) is the myogenic basic helix-loop-helix protein MyoD, discovered by Lassar and Weintraub [38]. Ectopic MyoD expression can convert mesenchymal stem cells or fibroblasts to a skeletal muscle cell fate [38]. Both of these examples are, in principle, consistent with the Waddington model of development, because the ectopic expression of a master regulatory factor in a developmentally "upstream" or multipotent cell type may simply cause a fate change at a point when the presumed epigenetic barriers are still relatively low. Indeed, more "downstream" (differentiated) cells are often more difficult to reprogram with MyoD [39]. However, the conversion of B-cells to a macrophage cell fate by forced expression of C/EBPα/β constitutes a clear example of the ability of these factors to convert one terminally differentiated somatic cell type into another [40].

In their seminal 2006 paper, Takahashi and Yamanaka identified a set of four transcription factors (Oct4, Sox2, Klf4, c-Myc) that is sufficient to reprogram mouse fibroblasts into induced pluripotent stem cells (iPSCs) [41]. It is a testament to the robustness of the process and the generalizability of the iPSC phenomenon that Yamanaka's somatic cell reprogramming technique was swiftly reproduced by many groups. The technique was shown to generate germline-competent mouse iPSCs efficiently and worked reliably with several somatic cell types or species, including human skin fibroblasts and blood cells [42–46]. Eventually, it became possible to generate pluripotent stem cells easily from any patient without the need to source human embryos or eggs.

2 Reprogramming Methods

Initially, all iPSCs were generated using integrating lenti/retroviral methods [43–45]. An inevitable drawback of this approach is that the proviral integration has unpredictable effects on the functional integrity or regulation of nearby genes [47]. In addition, the stably integrated reprogramming factor transgene expression cassettes are often incompletely silenced and can become re-activated during differentiation. Residual transgene expression can affect the performance and behavior of human iPSCs [48–51]. Mouse studies have shown that re-activation of the c-Myc transgene is associated with a high incidence of tumor formation in iPSC-chimeric mice [42, 52], making this method ill-suited for reliable disease modeling and potential clinical applications. Recombinase-mediated excision of the proviral reprogramming factor transgene cassette(s) is possible [48, 49], but this procedure is laborious and can alter the genome in an unpredictable manner.

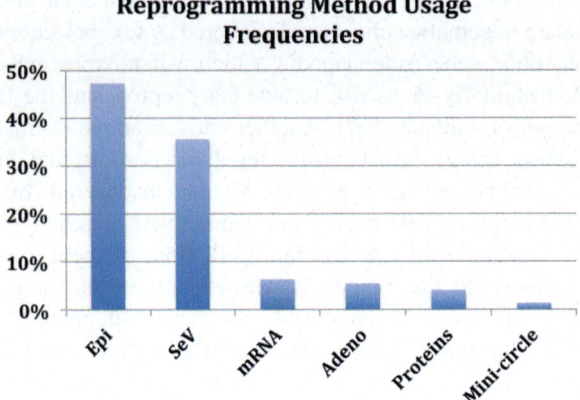

Fig. 1 Reprogramming method usage frequencies. PubMed searches were performed (December 2016) to estimate reprogramming method usage frequencies based on the number of PubMed indexed nonreview hiPSC publications containing identifying method-specific keywords in the title or abstract (actual method use was not confirmed)

These concerns have prompted the development of several alternative reprogramming methods that avoid stable integration. hiPSCs have been generated successfully by transiently exposing somatic cells to reprogramming factors through a variety of methods, including Sendai viral [53] or adenoviral [54] gene transfer, protein transduction [55], and transfection of mRNAs [56], microRNAs [57], minicircle [58] or episomally replicating plasmid DNA [59]. However, not all of these approaches have been widely adopted. A recent search of the current literature suggests that episomal- and Sendai virus-based techniques clearly dominate, with mRNA transfection-based methods a distant third (see Fig. 1). Although extremely high reprogramming efficiencies can be achieved with commercially available mRNA/microRNA transfection reprogramming kits, the higher workload, inability to reprogram hematopoietic cells, and lower reliability probably contributed to their relatively low adoption rate [60]. Therefore, the remainder of this review is focused on episomal and Sendai viral reprogramming modalities.

3   Nonintegrating Reprogramming with Episomally Replicating Plasmids

When conventional plasmids are transfected into human cells using methods such as electroporation, lipofection, or nucleofection, anywhere from <100 to >100,000 plasmid molecules enter each transfected cell [61–64]. Upon arriving in a transfected cell's nucleus, the plasmid-encoded genes are transcribed by the cell's transcriptional machinery. However, conventional plasmids lack an origin of DNA replication that is active in mammalian cells and, therefore, are not duplicated during cell division, resulting in a rapid decline in the number of episomal (i.e., extra-chromosomal, not integrated) plasmids per cell. Furthermore, plasmids are often recognized by transfected cells as foreign because of the presence of atypical or methylation-prone DNA sequences or a lack of associated chromatin factors and epigenetic marks [65–67]. Consequently, transiently transfected plasmids are quickly silenced or diluted out of the transfected cell population, and the short burst of plasmid gene expression is generally insufficient to reprogram somatic cells to pluripotency because reprogramming requires the transgenic reprogramming factors to be expressed for 8 days or longer [68–70]. Nevertheless, performing multiple rounds of plasmid transfection has led to the successful generation of hiPSCs, albeit at extremely low efficiencies [58, 71].

The ability to undergo DNA replication in proliferating mammalian cells can be conferred to plasmids by adding DNA virus-derived sequences that mediate viral DNA replication. These cis- and trans-acting elements interact with the cellular DNA replication machinery to facilitate duplication of the episomal virus DNA in actively cycling cells [72]. Epstein Barr virus (EBV)-derived episomal plasmids contain the EBV origin of replication (oriP) and express the Epstein Barr virus nuclear antigen 1 (EBNA1) protein that binds to the oriP, thereby tethering the plasmid to a chromosomal site and enabling replication of the plasmid in its episomal state [73]. However, for reasons that are not completely understood, only a small fraction (1–10%) of freshly transfected cells begin to replicate and maintain EBV-derived episomal plasmids [74]. In cells that do manage to maintain episomes, the plasmids persist at 100 or less copies per cell [75, 73]. Because each plasmid is replicated, at most, once per cell cycle [75], there is never any net increase in the number of episomal plasmids per cell. In fact, these episomal plasmids are eventually silenced or diluted out for a number of reasons, including inefficient plasmid replication [73, 74], uneven distribution to daughter cells [73], and DNA methylation [76].

Episomal reprogramming was first reported by the Thomson group in 2009 [77]. Since then, several groups have developed variations and improved versions of this approach (see Table 1). Most of these systems combine the conventional Yamanaka factors with additional reprogramming factors to increase reprogramming efficiencies, with each system using a specific combination of reprogramming factors and promoters. A frequently targeted additional pathway is the P53 tumor suppressor and genome stability gatekeeper pathway, a well-known bottleneck in retroviral somatic cell reprogramming ([78] and references therein). Several studies have explored whether inhibiting this pathway has beneficial effects in the context of episomal reprogramming. Indeed, increased episomal reprogramming efficiencies were observed when the P53 pathway was inactivated using a number of independent strategies, including coexpression of p53 shRNA [59, 79], dominant-negative TP53 [80], or SV40 large-T [77, 79].

Another useful auxiliary factor is BCL-XL. Forced expression of this protein, which has anti-apoptotic activities in blood cells and hPSCs [81, 82], seems to be especially useful in the context of episomal reprogramming of peripheral blood

Table 1 Key features of various episomal reprogramming systems

Laboratory	Thomson	Wang	Thermo	Muotri	Yamanaka	Zhang	Zhang	Yamanaka	Grzela	Thermo	Chen	Wu	Xu
PMID/Cat#	19325077	22132178	A14703	19763270	21460823	23704989	27161365	23193063	26088261	A15960	21243013	25628230	21399616
Promoter	EF1a, CMV	EF1a	EF1a	CMV	CAG	SFFV	SFFV	CAG	CAG	CAG	CAG	SFFV	CMV
OCT4	•	•	•	•	•	•	•	•	•	•	•	•	+VP16
SOX2	•	•	•	•	•	•	•	•	•	•	•	•	+VP16
KLF4	•	•	•	•	•	•	•	•	•	•	•	•	•
MYC	•	L-MYC	L-MYC		L-MYC	Optional	•	L-MYC	L-MYC	L-MYC	•	•	+VP16
NANOG	•	•	•							•	•	Optional	
LIN28	•	•			•			•		•		Optional	
p53 inactivation	LT	LT	LT		shRNA			shRNA	p53DD	p53DD	shRNA + LT	shRNA	
BCL-XL				•		•	•						
miRNA302/367	26584543								•	•			
Extra EBNA1	21478862							•	•				
Donor species	Human	Human	Human	Human	Human	Human	Human	Human	Human	Human	Human	Mouse + human	Mouse only

mononuclear cells (PBMCs) ([83, 84, 85] and our unpublished observation). Furthermore, several groups have shown that transiently augmenting EBNA1 expression levels at the beginning of reprogramming elevates episomal reprogramming efficiencies ([86, 80, 87] and our unpublished observation), presumably by aiding episomal plasmid replication during the first cell divisions, although EBNA1 expression can have growth-promoting effects on its own [88]. Members of the microRNA cluster 302/367 have been shown to increase human fibroblast reprogramming efficiencies in lentiviral reprogramming studies [89]. These pluripotency-associated microRNAs have many targets, including genes involved in the regulation of cell proliferation, apoptosis, and mesenchymal-to-epithelial transition [90, 91]. Several groups have shown that inclusion of these microRNAs promotes episomal reprogramming of human fibroblasts [92, 86, 80]. However, the reprogramming-boosting effect of the cluster 302/367 microRNAs does not appear to extend to hematopoietic progenitor cell reprogramming ([86] and our unpublished observation).

A systematic comparison of the efficiency and reliability of the episomal reprogramming systems shown in Table 1 has not been published, and the number of independent studies that use the same system to reprogram the same somatic cell type is generally quite low. One of the many factors that confound efficiency comparison across studies is that different somatic cell types often reprogram at vastly different rates. For example, the original Thomson system seems to work much more efficiently with the nonlymphocytic peripheral blood mononuclear cell compartment than with fibroblasts [93]. Significant cell type-dependent efficiency differences have also been reported for the Yamanaka episomal system, with 500 times more hiPSC colonies emerging when urinary epithelial cells are reprogrammed compared with scar tissue fibroblasts [80]. Moreover, even when the same laboratory applies the same episomal reprogramming to different specimens of the same cell type, the reprogramming efficiencies can vary 100-fold [60]. Efficiency comparison analyses are often confounded even further by the use of different markers or standards to identify and enumerate emerging hiPSCs [70, 60], by the variable amount of proliferation of different somatic cell population [86], and by differences in cell plating densities or the number of replatings (passaging cells during reprogramming can result in multiple hiPSC colonies emerging from the same original somatic cell, thus leading to artificially inflated efficiencies). Thus, it is difficult to draw firm conclusions about the relative strengths or weaknesses of any one system. Nevertheless, when significant observations are reported consistently by independent studies they are more likely to reflect true biological differences. For instance, the increased efficiency that was realized by Yamanaka's modifications to the original (Thomson) episomal reprogramming platform ([86, 59]; see Table 1) was confirmed by Goh et al. [94], and we have also observed this phenomenon (unpublished observation). On the other hand, a version of the Thomson system in which L-Myc is used instead of c-Myc was shown to work well with freshly isolated blood cells from multiple patients when a cocktail of small-molecule kinase inhibitors was included [95]. Improved versions of both systems are available as kits sold by Thermo (see Table 1).

The fact that only a very small number of all possible reprogramming factor combinations have been tested so far provides some hope that much more efficient episomal reprogramming factor cocktails may yet be identified. Particularly promising would be factors that help initiate episomal plasmid replication in transfected cells, because such factors could result in more than tenfold increase in efficiency (see above).

The episomal approach to reprogramming is a particularly attractive test bed because of the simplicity of the agent and the ease of plasmid construction and production relative to virus-based systems. A promising approach, besides the testing of new factors, is the augmentation of proven reprogramming factors with the addition of a strong, heterologous trans-activating domain (TAD) that can substantially augment the activity of transcription factors [96]. Expression of an OCT4-VP16 TAD fusion gene was sufficient to reprogram mouse embryonic fibroblasts into high-quality iPS cells [97]. The addition of a TAD also enhanced the ability of OCT4 to reprogram human somatic cells [97, 98]. However, reprogramming factors must mediate not only the re-activation of pluripotency genes but also the silencing of somatically expressed genes. Indeed, the core triumvirate of pluripotency factors (OCT4, SOX2, NANOG) occupies both active and inactive genes in hPSCs [99, 100]. Thus, it is surprising that the addition of a strong TAD enhances the reprogramming activity of OCT4 or SOX2 [101]. Indeed, KLF4 primarily acts as a transcriptional repressor, and its reprogramming activity was not enhanced by the addition of a TAD [97, 98]. For reasons that are not entirely clear, OCT4 could only be enhanced by some TADs (MyoD, VP16, YAP) and not others [97, 101, 102]. The use of TADs clearly has great potential, yet this strategy has not yet been fully explored in the context of episomal reprogramming (see Table 1).

The use of small molecules in reprogramming to pluripotency constitutes another opportunity to make reprogramming more efficient and reliable, but adds yet another level of complexity. Commonly used inhibitors target MEK (e.g., PD0325901), GSKβ (CHIR99021), TGFβ (A-83-01 or SB-431542), ROCK/myosin (HA-100, Y-27632, blebbistatin), and HDAC (butyrate, VPA) pathways (an - in-depth discussion of reprogramming-enhancing small molecules is beyond the scope of this article but several reviews have been published recently [103, 104]).

A key advantage of episomal reprogramming over lenti/retroviral reprogramming is that the nonintegrating episomal approach promises to yield transgene-free and genomically intact hiPSCs. Several studies have shown that the majority of episomally derived hiPSC lines do indeed become devoid of detectable plasmid DNA sequences by passage 5–10 [59, 77, 60], even when additional plasmids are used to boost the initial levels of EBNA1 [86]. Nevertheless, a major concern with episomal reprogramming is the risk that exogenous plasmid DNAs (or fragments thereof) persist in these hiPSCs, either through continued episomal replication or as a consequence of stable chromosomal integration, thus potentially causing insertional mutagenesis or leading to constitutive expression of proto-oncogenes or P53 pathway antagonists. Absence of exogenous plasmid-derived DNA sequences is therefore an important criterion of high-quality episomal hiPSCs. However, because it is generally difficult to prove a negative, the level of scrutiny needs to be carefully balanced with the perceived risk and potential damage resulting from persistent exogenous DNA sequences. If, for example, the hiPSCs are only used for in vitro research studies, it may be sufficient to produce two or three independent hiPSC lines that remain quantitative polymerase chain reaction (qPCR)-negative for EBNA1 DNA sequences [60]. If, on the other hand, an episomal hiPSC line is being evaluated for use in manufacturing cell-based therapeutics for the treatment of the somatic cell donor, the safety concerns are much more significant, and can increase even further if the hiPSCs are to be used in the large-scale manufacture of allogeneic cell therapeutics. In these cases, absence of any plasmid DNA sequences must be shown by several independent, carefully validated, and highly sensitive and accurate methods, such as Droplet Digital (dd) PCR (looking at multiple exogenous target sequences including EBNA1 and the reprogramming factor ORFs), as well as by whole genome sequencing at multifold coverage. For all of these analyses it is important to prepare the DNA in a manner that captures both genomic and extrachromosomal plasmid DNA at high efficiencies (e.g., using a direct lysis method [60]). At the same time, extreme care must be taken to avoid false-positives resulting from cross-contamination with plasmid DNAs that are handled in nearby laboratory spaces. Assuming a quality threshold of <1 plasmid detected per 1,000 cells, a seemingly small amount of plasmid DNA (e.g., 1 ng or 1-millionth of a typical plasmid maxi-preparation) could render ~10^{11} cells (the equivalent of ~10,000 flasks) false-positive.

Among the advantages of episomal reprogramming is the diversity of cell types that can be successfully reprogrammed, including dermal fibroblasts, blood cells, mesenchymal stem cells, and urinary cells [80, 86, 105, 59]. Furthermore, the simplicity of the reagent (plasmid DNA), the high reliability of the method, and the availability of clinical-grade episomal reprogramming protocols make episomal reprogramming the method of choice for the production of clinical-grade hiPSCs (e.g., [60, 106, 107, 87, 94]). The relatively low cost of the reagent contributes to the popularity of research-grade reprogramming using episomal vectors (see Fig. 1 and [60]).

A potential downside of episomal reprogramming systems is the requirement for expensive transfection equipment, such as the Lonza/Amaxa 2D/4D nucleofector™ or the Thermo Neon™ device. A more serious concern is the genetic integrity of the produced hiPSCs. In addition to the already discussed potential for persistence of episomal plasmid-derived DNA, we observed a slightly increased rate of genetic abnormalities [60] that may be even more pronounced when chemically defined media and matrix reagents are used ([87], and our unpublished observation). To be clear, only a minority of episomal hiPSCs are affected by gross abnormalities that can be detected by conventional karyotyping, and this frequency is still below the average frequency reported in a large cross-laboratory meta-analysis of hESC and hiPSC karyotypes [108]. Nevertheless, the frequency was somewhat higher than for Sendai viral or mRNA transfection-based reprogramming performed in the same laboratory [60]. To generate the highest quality episomal hiPSCs, the inclusion of a subcloning step is recommended. Several independent (i.e., derived from distinct somatic cells) subclonal hiPSC lines should be isolated and tested to find at least two to three lines that meet the predetermined safety/quality criteria. As is the case for any hPSC culture, it is important to repeat the genetic integrity tests at regular intervals because genetic mutations can occur at any time and such alterations can generate a growth advantage that allows a single mutant cell's progeny to increase exponentially in frequency. Growth advantages can result from the presence of transgenes or the mutation of endogenous genes involved in cell growth or survival.

4   Sendai-Viral Reprogramming

The other widely practiced reprogramming method employs Sendai virus-derived particles. Sendai virus (SeV, murine parainfluenza virus type 1) is an RNA virus that mostly affects respiratory tissues in rodents [109]. It binds to sialic acid residues expressed on target cells, allowing it to infect many murine and human cell types readily. Unlike its human counterpart (human parainfluenza virus), SeV is generally not pathogenic in humans and, although able to efficiently infect human cells, fails to evade the human innate immune system [110]. The SeV envelope comprises a host-cell-derived membrane that contains viral proteins HN and F, which are involved in virus attachment and membrane fusion/viral entry, respectively. Located underneath the envelope is a structural matrix protein called M that is important for viral assembly and budding. The single-stranded genomic RNA molecule, over 15 kb in size, exists as a ribonucleoprotein complex that chiefly contains nucleocapsid protein (NP) molecules. SeV RNA replication and amplification, which typically creates over 10^4 copies per cell [111], occurs in the cytoplasm of infected cells where the viral RNA-dependent RNA polymerase (L and P proteins) replicates the NP-bound viral RNA genome and also produces translatable positive-strand RNAs for protein expression. In brief, proteins NP, L, and P are crucial for SeV RNA replication and gene expression whereas HN, M, and F mediate attachment and membrane fusion during cell entry as well as budding and release of new virions.

The Hasegawa group was the first to report the successful reprogramming of human fibroblasts using SeV-based transfer of reprogramming factors [53]. Rather than trying to package all four classical Yamanaka factor genes (Oct4, Sox2, Klf4, c-Myc; often written as OSKM) into a single virus, this group generated separate SeV constructs for each reprogramming factor gene. Reasons against the all-in-one strategy included (discussed in [112]) (1) the probable increase in tumorigenicity for all-in-one SeVs; (2) concerns about a drop in titer due to the significantly larger insert size; (3) potential detrimental effects of coexpressing OSKM in SeV producer cells; and (4) reduced control over reprogramming factor stoichiometry with an all-in-one system. A standard safety feature of SeV vectors is deletion of the F gene from the viral genome, which renders SeV particles produced by infected cells incapable of infecting other cells [113]. To limit the errant production of nontransmissible virus-like particles, temperature-sensitive versions [114] of the

M and HN genes were used and the viruses generated at a permissible (low) temperature. Additional mutations were introduced into the viral RNA polymerase subunit genes P and L [112, 115] to boost long-term replication and transgene expression. When the Hasegawa group cotransduced human fibroblasts with these SeVs at a multiplicity of infection (MOI) of 3, up to 1% of the cells turned into integration-free hiPSCs. However, viral replication was so efficient that, in several of these hiPSC lines, transgenic RNA could still be detected at passage 15 and beyond [53, 112]. In a follow-up study, the system was improved by including additional temperature-sensitive mutations into the P and L genes of the c-MYC SeV [116]. Once the conventional SeV constructs carrying the other three reprogramming factors and the more active versions of the P and L genes had disappeared from the cells, these additional mutations strongly reduced SeV RNA expression and replication. This system was commercialized and sold as a human somatic cell reprogramming kit (CytoTune™).

The efficiency of hiPSC production and the ability to remove persistent SeV RNAs were further increased by coexpressing KLF4, OCT4, and SOX2 from a single SeV construct that also carried the additional P gene mutations [117]. The ability to increase SeV RNA polymerase activity in transduced cells by adding the original KLF4 SeV (carrying the more active versions of the P and L genes) makes this system quite flexible. If reprogramming efficiencies are too low with a particular patient sample or somatic cell type, more of the KLF4 SeV can be added to increase replication of all viruses during the early phase of reprogramming. Conversely, if persistence of SeV RNAs or high levels of cytotoxicity are observed, the MOI of some or all of the SeVs can be reduced. This system, now sold as the CytoTune™ 2.0 kit, has higher efficiency, lower cytotoxicity, and faster viral clearance than the original kit [118, 60]. Cell types that have been reprogrammed with CytoTune™ kits include fibroblasts, PBMCs (including T-cells), and keratinocytes [119].

The Nakanishi group [120] also employed altered versions of the P and L proteins to boost transgene expression and limit activation of the interferon pathway and the resulting cytopathic effect that is often triggered by wild-type SeV [121]. Deletion of the entire M, F, and HN gene region removed their toxic effects and generated enough space to accommodate the genes for all four reprogramming factors, thereby ensuring that each transduced cell received a complete set. This approach probably results in a more precise reprogramming factor stoichiometry compared with cotransduction with separate SeVs that carry individual factors [120, 53]. However, whether these features actually result in higher reprogramming efficiencies and more completely reprogrammed hiPSCs remains to be seen because this system is not widely available and has so far only been used to create mouse iPSCs. Viral RNA replication was very efficient for the reprogramming of mouse cells with this system and continued in the iPSCs. Nevertheless, SeV RNA-free mouse iPSCs could be obtained after transiently transfecting cells with siRNAs directed against the L gene. The key features of all of these systems are summarized in Table 2.

Table 2 Key features of various Sendai viral reprogramming systems

Reference	Fusaki et al. [53], Fusaki and Ban [112]		Ban et al. [116], Cytotune™		Fujie et al. [117]		Cytotune 2™			Nishimura et al. [120]
Transgene	O.S.K	M	O.S.K	M	KOS	M	KOS	K	M	MKOS
Position	Ie/N	HN/L	Ie/N	HN/L	P/M	HN/L	P/M	Ie/N	HN/L	(P/L)
SeV protein mutations — M	G69E T116A A183S	G69E T116A A183S	G69E T116A A183S	G69E T116A A183S	G69E T116A A183S	G69E T116A A183S	G69E T116A A183S	G69E T116A A183S	G69E T116A A183S	Δ
HN	A262T G264R K461G	A262T G264R K461G	A262T G264R K461G	A262T G264R K461G	A262T G264R K461G	A262T G264R K461G	A262T G264R K461G	A262T G264R K461G	A262T G264R K461G	Δ
F	Δ	Δ	Δ	Δ	Δ	Δ	Δ	Δ	Δ	Δ
P	L511F	L511F	L511F	D433A R434A K437A L511F	D433A R434A K437A L511F	D433A R434A K437A L511F	D433A R434A K437A L511F	L511F	D433A R434A K437A L511F	P517H
L	N1197S K1795E	N1197S K1795E	N1197S K1795E	N1197S L1361C L1558I K1795E	N1197S K1795E	N1197S L1361C L1558I K1795E	N1197S K1795E	N1197S K1795E	N1197S L1361C L1558I K1795E	I981V A1088S S1207C L1618V
Replication — 35°C	+++	+++	+++	++	+++	++	+++	+++	++	+++
37°C	+++	+++	+++	(+)	++	(+)	++	+++	(+)	+++
39°C	–	–	–	–	–	–	–	–	–	TBD
Comment		Often presists in hiPSCs for many passages		Efficient replication at 37°C requires continuous complementation by the other SeVs	Reduced ability to replicate at 37°C. Three-in-one design increases reprogramming efficiency	Efficient replication at 37°C requires continuous complementation by the other SeVs	Reduced ability to replicate at 37°C. Three-in-one design increases reprogramming efficiency	Original (Fusaki) SeV for KLF4. Strong ability to replicate, can assist replication of KOS and M SeVs in trans	Efficient replication at 37°C requires continuous complementation by the other SeVs	Strong replication (removal requires anti-L-siRNA treatment). High efficiency, absence of non-permissible particles. Only mouse iPSCs have been reported

5 Episomal or Sendai-Viral Reprogramming?

It is now well established that both the episomal and the Sendai viral approaches to reprogramming work reliably, and many studies have tried and (mostly) failed to find consistent and meaningful differences between the hiPSCs created using these methods [60, 122, 123], or between hiPSCs and hESC in general [124–127, 60, 122], although some subtle, method-specific differences may exist [128, 60, 129]. Episomal reprogramming is currently the preferred method for creating clinical-grade hiPSCs [84, 60, 87, 107, 94], including the first-in-human trial involving hiPSCs [130], although this preference is not absolute [131, 132] and current good manufacturing practice (cGMP)-grade SeVs for reprogramming have become available from DNAvec/ID-Pharma. Episomal reprogramming has a clear cost advantage, especially if a suitable transfection device is already available to the user and when the plasmids are prepared in-house. Other advantages of the episomal system include the lack of the cytopathic effect that is observed with SeV reprogramming ([118] and our unpublished observation) as well as the ease with which newly identified reprogramming factors can be added and tested. The differences in reprogramming efficiencies between episomal and SeV reprogramming are probably mainly caused by the relatively low percentage of somatic cells that successfully establish replication of episomal plasmids. The difference in the number of colonies produced for each somatic cell can be substantial [60]. However, in most reprogramming scenarios, the higher efficiency of SeV reprogramming simply translates into an even larger surplus of colonies [60] and, therefore, is not of great practical significance. SeV reprogramming with the CytoTune™ kit is very easy but expensive if only a one samples is reprogrammed per single-use kit. On the other hand, the very high reliability and efficiency of SeV reprogramming often allows many samples to be reprogrammed with a single kit, either in parallel or by re-freezing leftover virus [118]. In summary, the choice between these two methods depends on laboratory- and project-specific preferences and circumstances [133, 134].

References

1. Birch S, Tyson E (1683) An extract of two letters from Mr. Sampson birch, an alderman and apothecary at stafford, concerning an extraordinary birth in Staffordshire, with reflections thereon by Edw. Tyson M. D. fellow of the coll. of physitians, and of the R. society. Philos Trans 13:281–284
2. Virchow R (1863) Die krankhaften Geschwülste. Hirschwald
3. Kleinsmith LJ, Pierce Jr GB (1964) Multipotentiality of single embryonal carcinoma cells. Cancer Res 24:1544–1551
4. Martin GR, Evans MJ (1975) Differentiation of clonal lines of teratocarcinoma cells: formation of embryoid bodies in vitro. Proc Natl Acad Sci U S A 72:1441–1445
5. Papaioannou VE, Gardner RL, McBurney MW, Babinet C, Evans MJ (1978) Participation of cultured teratocarcinoma cells in mouse embryogenesis. J Embryol Exp Morphol 44:93–104

6. Rossant J, McBurney MW (1982) The developmental potential of a euploid male teratocarcinoma cell line after blastocyst injection. J Embryol Exp Morphol 70:99–112

7. Papaioannou VE, McBurney MW, Gardner RL, Evans MJ (1975) Fate of teratocarcinoma cells injected into early mouse embryos. Nature 258:70–73

8. Brinster RL (1974) The effect of cells transferred into the mouse blastocyst on subsequent development. J Exp Med 140:1049–1056

9. Askanazy M (1907) Die Teratome nach ihrem Bau, ihrem Verlauf, ihrer Genese und im Vergleich zum experimentellen Teratoid. Verh Deutsch Pathol Gesellsch 11:39–82

10. Stevens LC (1970) The development of transplantable teratocarcinomas from intratesticular grafts of pre- and postimplantation mouse embryos. Dev Biol 21:364–382

11. Solter D, Skreb N, Damjanov I (1970) Extrauterine growth of mouse egg-cylinders results in malignant teratoma. Nature 227:503–504

12. Evans MJ, Kaufman MH (1981) Establishment in culture of pluripotential cells from mouse embryos. Nature 292:154–156

13. Martin GR (1981) Isolation of a pluripotent cell line from early mouse embryos cultured in medium conditioned by teratocarcinoma stem cells. Proc Natl Acad Sci U S A 78:7634–7638

14. Bradley A, Evans M, Kaufman MH, Robertson E (1984) Formation of germ-line chimaeras from embryo-derived teratocarcinoma cell lines. Nature 309:255–256

15. Nagy A et al (1990) Embryonic stem cells alone are able to support fetal development in the mouse. Development 110:815–821

16. Doetschman T et al (1987) Targetted correction of a mutant HPRT gene in mouse embryonic stem cells. Nature 330:576–578

17. Thomas KR, Capecchi MR (1987) Site-directed mutagenesis by gene targeting in mouse embryo-derived stem cells. Cell 51:503–512

18. National Research Council (1994) Genetically altered mice: a revolutionary research resource. In: Sharing laboratory resources: genetically altered mice: summary of a workshop held at the National Academy of Sciences, 23–24 March 1993. https://www.ncbi.nlm.nih.gov/books/NBK231336/

19. Bongso A, Fong CY, Ng SC, Ratnam S (1994) Isolation and culture of inner cell mass cells from human blastocysts. Hum Reprod 9:2110–2117

20. Thomson JA et al (1998) Embryonic stem cell lines derived from human blastocysts. Science 282:1145–1147

21. Tesar PJ et al (2007) New cell lines from mouse epiblast share defining features with human embryonic stem cells. Nature 448:196–199

22. Brons IG et al (2007) Derivation of pluripotent epiblast stem cells from mammalian embryos. Nature 448:191–195

23. Batlle-Morera L, Smith A, Nichols J (2008) Parameters influencing derivation of embryonic stem cells from murine embryos. Genesis 46:758–767

24. Tachibana M et al (2013) Human embryonic stem cells derived by somatic cell nuclear transfer. Cell 153:1228–1238

25. Oderberg DS (2005) Human embryonic stem cell research: what's wrong with it? Hum Life Rev 31:21–33

26. De Los Angeles A et al (2015) Hallmarks of pluripotency. Nature 525:469–478

27. Mascetti VL, Pedersen RA (2016) Contributions of mammalian chimeras to pluripotent stem cell research. Cell Stem Cell 19:163–175

28. Muller FJ, Brandl B, Loring JF (2008) Assessment of human pluripotent stem cells with PluriTest. StemBook 2012. Harvard Stem Cell Institute, Cambridge

29. Waddington CH (1957) The strategy of the genes. Routledge, Abingdon

30. Gurdon JB (1962) The developmental capacity of nuclei taken from intestinal epithelium cells of feeding tadpoles. J Embryol Exp Morphol 10:622–640

31. Campbell KH, McWhir J, Ritchie WA, Wilmut I (1996) Sheep cloned by nuclear transfer from a cultured cell line. Nature 380:64–66

32. Miller RA, Ruddle FH (1976) Pluripotent teratocarcinoma-thymus somatic cell hybrids. Cell 9:45–55
33. Ying QL, Nichols J, Evans EP, Smith AG (2002) Changing potency by spontaneous fusion. Nature 416:545–548
34. Terada N et al (2002) Bone marrow cells adopt the phenotype of other cells by spontaneous cell fusion. Nature 416:542–545
35. Do JT, Scholer HR (2004) Nuclei of embryonic stem cells reprogram somatic cells. Stem Cells 22:941–949
36. Cowan CA, Atienza J, Melton DA, Eggan K (2005) Nuclear reprogramming of somatic cells after fusion with human embryonic stem cells. Science 309:1369–1373
37. Schneuwly S, Klemenz R, Gehring WJ (1987) Redesigning the body plan of drosophila by ectopic expression of the homoeotic gene antennapedia. Nature 325:816–818
38. Davis RL, Weintraub H, Lassar AB (1987) Expression of a single transfected cDNA converts fibroblasts to myoblasts. Cell 51:987–1000
39. Filvaroff EH, Derynck R (1996) Induction of myogenesis in mesenchymal cells by MyoD depends on their degree of differentiation. Dev Biol 178:459–471
40. Xie H, Ye M, Feng R, Graf T (2004) Stepwise reprogramming of B cells into macrophages. Cell 117:663–676
41. Takahashi K, Yamanaka S (2006) Induction of pluripotent stem cells from mouse embryonic and adult fibroblast cultures by defined factors. Cell 126:663–676
42. Okita K, Ichisaka T, Yamanaka S (2007) Generation of germline-competent induced pluripotent stem cells. Nature 448:313–317
43. Yu J et al (2007) Induced pluripotent stem cell lines derived from human somatic cells. Science 318:1917–1920
44. Takahashi K et al (2007) Induction of pluripotent stem cells from adult human fibroblasts by defined factors. Cell 131:861–872
45. Park IH et al (2008) Reprogramming of human somatic cells to pluripotency with defined factors. Nature 451:141–146
46. Loh YH et al (2010) Reprogramming of T cells from human peripheral blood. Cell Stem Cell 7:15–19
47. Hacein-Bey-Abina S et al (2003) LMO2-associated clonal T cell proliferation in two patients after gene therapy for SCID-X1. Science 302:415–419
48. Soldner F et al (2009) Parkinson's disease patient-derived induced pluripotent stem cells free of viral reprogramming factors. Cell 136:964–977
49. Sommer CA et al (2010) Excision of reprogramming transgenes improves the differentiation potential of iPS cells generated with a single excisable vector. Stem Cells 28:64–74
50. Ramos-Mejia V et al (2012) Residual expression of the reprogramming factors prevents differentiation of iPSC generated from human fibroblasts and cord blood CD34+ progenitors. PLoS One 7:e35824
51. Awe JP et al (2013) Generation and characterization of transgene-free human induced pluripotent stem cells and conversion to putative clinical-grade status. Stem Cell Res Ther 7:87
52. Nakagawa M et al (2008) Generation of induced pluripotent stem cells without Myc from mouse and human fibroblasts. Nat Biotechnol 26:101–106
53. Fusaki N, Ban H, Nishiyama A, Saeki K, Hasegawa M (2009) Efficient induction of transgene-free human pluripotent stem cells using a vector based on Sendai virus, an RNA virus that does not integrate into the host genome. Proc Jpn Acad Ser B Phys Biol Sci 85:348–362
54. Zhou W et al (2009) Adenoviral gene delivery can reprogram human fibroblasts to induced pluripotent stem cells. Stem Cells 27:2667–2674
55. Kim D et al (2009) Generation of human induced pluripotent stem cells by direct delivery of reprogramming proteins. Cell Stem Cell 4:472–476
56. Warren L et al (2010) Highly efficient reprogramming to pluripotency and directed differentiation of human cells with synthetic modified mRNA. Cell Stem Cell 7:618–630

57. Miyoshi N et al (2011) Reprogramming of mouse and human cells to pluripotency using mature microRNAs. Cell Stem Cell 8:633–638
58. Jia F et al (2010) A nonviral minicircle vector for deriving human iPS cells. Nat Methods 7:197–199
59. Okita K et al (2011) A more efficient method to generate integration-free human iPS cells. Nat Methods 8:409–412
60. Schlaeger TM et al (2015) A comparison of non-integrating reprogramming methods. Nat Biotechnol 33:58–63
61. Carapuca E, Azzoni AR, Prazeres DM, Monteiro GA, Mergulhao FJ (2007) Time-course determination of plasmid content in eukaryotic and prokaryotic cells using real-time PCR. Mol Biotechnol 37:120–126
62. Gershan JA et al (2005) Immediate transfection of patient-derived leukemia: a novel source for generating cell-based vaccines. Genet Vaccines Ther 3:4
63. Cohen RN, van der Aa MA, Macaraeg N, Lee AP, Szoka Jr FC (2009) Quantification of plasmid DNA copies in the nucleus after lipoplex and polyplex transfection. J Control Release 135:166–174
64. Fliedl L, Kast F, Grillari J, Wieser M, Grillari-Voglauer R (2015) Optimization of a quantitative PCR based method for plasmid copy number determination in human cell lines. New Biotechnol 32:716–719
65. Chen ZY, He CY, Ehrhardt A, Kay MA (2003) Minicircle DNA vectors devoid of bacterial DNA result in persistent and high-level transgene expression in vivo. Mol Ther 8:495–500
66. Riu E, Chen ZY, Xu H, He CY, Kay MA (2007) Histone modifications are associated with the persistence or silencing of vector-mediated transgene expression in vivo. Mol Ther 15:1348–1355
67. Gracey Maniar LE et al (2013) Minicircle DNA vectors achieve sustained expression reflected by active chromatin and transcriptional level. Mol Ther 21:131–138
68. Maherali N et al (2008) A high-efficiency system for the generation and study of human induced pluripotent stem cells. Cell Stem Cell 3:340–345
69. Hockemeyer D et al (2008) A drug-inducible system for direct reprogramming of human somatic cells to pluripotency. Cell Stem Cell 3:346–353
70. Chan EM et al (2009) Live cell imaging distinguishes bona fide human iPS cells from partially reprogrammed cells. Nat Biotechnol 27:1033–1037
71. Si-Tayeb K et al (2010) Generation of human induced pluripotent stem cells by simple transient transfection of plasmid DNA encoding reprogramming factors. BMC Dev Biol 10:81
72. Van Craenenbroeck K, Vanhoenacker P, Haegeman G (2000) Episomal vectors for gene expression in mammalian cells. Eur J Biochem 267:5665–5678
73. Nanbo A, Sugden A, Sugden B (2007) The coupling of synthesis and partitioning of EBV's plasmid replicon is revealed in live cells. EMBO J 26:4252–4262
74. Leight ER, Sugden B (2001) Establishment of an oriP replicon is dependent upon an infrequent, epigenetic event. Mol Cell Biol 21:4149–4161
75. Yates JL, Guan N (1991) Epstein-Barr virus-derived plasmids replicate only once per cell cycle and are not amplified after entry into cells. J Virol 65:483–488
76. Kameda T, Smuga-Otto K, Thomson JA (2006) A severe de novo methylation of episomal vectors by human ES cells. Biochem Biophys Res Commun 349:1269–1277
77. Yu J et al (2009) Human induced pluripotent stem cells free of vector and transgene sequences. Science 324:797–801
78. Krizhanovsky V, Lowe SW (2009) Stem cells: the promises and perils of p53. Nature 460:1085–1086
79. Chou BK et al (2011) Efficient human iPS cell derivation by a non-integrating plasmid from blood cells with unique epigenetic and gene expression signatures. Cell Res 21:518–529
80. Drozd AM et al (2015) Generation of human iPSCs from cells of fibroblastic and epithelial origin by means of the oriP/EBNA-1 episomal reprogramming system. Stem Cell Res Ther 6:122

81. Silva M et al (1996) Erythropoietin can promote erythroid progenitor survival by repressing apoptosis through Bcl-XL and Bcl-2. Blood 88:1576–1582
82. Bai H et al (2012) Bcl-xL enhances single-cell survival and expansion of human embryonic stem cells without affecting self-renewal. Stem Cell Res 8:26–37
83. Su RJ et al (2013) Efficient generation of integration-free ips cells from human adult peripheral blood using BCL-XL together with Yamanaka factors. PLoS One 8:e64496
84. Chou BK et al (2015) A facile method to establish human induced pluripotent stem cells from adult blood cells under feeder-free and xeno-free culture conditions: a clinically compliant approach. Stem Cells Transl Med 4:320–332
85. Wen W et al (2016) Enhanced generation of integration-free iPSCs from human adult peripheral blood mononuclear cells with an optimal combination of episomal vectors. Stem Cell Rep 6:873–884
86. Okita K et al (2013) An efficient nonviral method to generate integration-free human-induced pluripotent stem cells from cord blood and peripheral blood cells. Stem Cells 31:458–466
87. Chen G et al (2011) Chemically defined conditions for human iPSC derivation and culture. Nat Methods 8:424–429
88. Frappier L (2012) Contributions of epstein-barr nuclear antigen 1 (EBNA1) to cell immortalization and survival. Virus 4:1537–1547
89. Anokye-Danso F et al (2011) Highly efficient miRNA-mediated reprogramming of mouse and human somatic cells to pluripotency. Cell Stem Cell 8:376–388
90. Zhang Z et al (2015) MicroRNA-302/367 cluster governs hESC self-renewal by dually regulating cell cycle and apoptosis pathways. Stem Cell Rep 4:645–657
91. Subramanyam D et al (2011) Multiple targets of miR-302 and miR-372 promote reprogramming of human fibroblasts to induced pluripotent stem cells. Nat Biotechnol 29:443–448
92. Howden SE et al (2015) Simultaneous reprogramming and gene correction of patient fibroblasts. Stem Cell Rep 5:1109–1118
93. Hu K et al (2011) Efficient generation of transgene-free induced pluripotent stem cells from normal and neoplastic bone marrow and cord blood mononuclear cells. Blood 117:e109–e119
94. Goh PA et al (2013) A systematic evaluation of integration free reprogramming methods for deriving clinically relevant patient specific induced pluripotent stem (iPS) cells. PLoS One 8: e81622
95. Mack AA, Kroboth S, Rajesh D, Wang WB (2011) Generation of induced pluripotent stem cells from CD34+ cells across blood drawn from multiple donors with non-integrating episomal vectors. PLoS One 6:e27956
96. Sadowski I, Ma J, Triezenberg S, Ptashne M (1988) GAL4-VP16 is an unusually potent transcriptional activator. Nature 335:563–564
97. Wang Y et al (2011) Reprogramming of mouse and human somatic cells by high-performance engineered factors. EMBO Rep 12:373–378
98. Hirai H et al (2011) Radical acceleration of nuclear reprogramming by chromatin remodeling with the transactivation domain of MyoD. Stem Cells 29:1349–1361
99. Boyer LA et al (2005) Core transcriptional regulatory circuitry in human embryonic stem cells. Cell 122:947–956
100. Schmidt R, Plath K (2012) The roles of the reprogramming factors Oct4, Sox2 and Klf4 in resetting the somatic cell epigenome during induced pluripotent stem cell generation. Genome Biol 13:251
101. Zhu G et al (2014) Coordination of engineered factors with TET1/2 promotes early-stage epigenetic modification during somatic cell reprogramming. Stem Cell Rep 2:253–261
102. Hirai H, Katoku-Kikyo N, Karian P, Firpo M, Kikyo N (2012) Efficient iPS cell production with the MyoD transactivation domain in serum-free culture. PLoS One 7:e34149
103. Federation AJ, Bradner JE, Meissner A (2014) The use of small molecules in somatic-cell reprogramming. Trends Cell Biol 24:179–187

104. Li W, Li K, Wei W, Ding S (2013) Chemical approaches to stem cell biology and therapeutics. Cell Stem Cell 13:270–283
105. Yan X et al (2016) Generation of induced pluripotent stem cells from human mesenchymal stem cells of parotid gland origin. Am J Transl Res 8:419–432
106. Baghbaderani BA et al (2016) Detailed characterization of human induced pluripotent stem cells manufactured for therapeutic applications. Stem Cell Rev 12:394–420
107. Baghbaderani BA et al (2015) cGMP-manufactured human induced pluripotent stem cells are available for pre-clinical and clinical applications. Stem Cell Rep 5:647–659
108. Taapken SM et al (2011) Karotypic abnormalities in human induced pluripotent stem cells and embryonic stem cells. Nat Biotechnol 29:313–314
109. Nagai Y (ed) (2013) Sendai virus vector. Springer, Tokyo
110. Bousse T, Chambers RL, Scroggs RA, Portner A, Takimoto T (2006) Human parainfluenza virus type 1 but not Sendai virus replicates in human respiratory cells despite IFN treatment. Virus Res 121:23–32
111. Nishimura K et al (2007) Persistent and stable gene expression by a cytoplasmic RNA replicon based on a noncytopathic variant Sendai virus. J Biol Chem 282:27383–27391
112. Fusaki N, Ban H (2013) Induction of human pluripotent stem cells by the Sendai virus vector: establishment of a highly efficient and footprint-free system. In: Nagai Y (ed) Sendai virus vector. Springer, Tokyo, pp 171–183
113. Li HO et al (2000) A cytoplasmic RNA vector derived from nontransmissible Sendai virus with efficient gene transfer and expression. J Virol 74:6564–6569
114. Inoue M et al (2003) Nontransmissible virus-like particle formation by F-deficient sendai virus is temperature sensitive and reduced by mutations in M and HN proteins. J Virol 77:3238–3246
115. Yoshizaki M et al (2006) Naked Sendai virus vector lacking all of the envelope-related genes: reduced cytopathogenicity and immunogenicity. J Gene Med 8:1151–1159
116. Ban H et al (2011) Efficient generation of transgene-free human induced pluripotent stem cells (iPSCs) by temperature-sensitive Sendai virus vectors. Proc Natl Acad Sci U S A 108:14234–14239
117. Fujie Y et al (2014) New type of Sendai virus vector provides transgene-free iPS cells derived from chimpanzee blood. PLoS One 9:e113052
118. Beers J et al (2015) A cost-effective and efficient reprogramming platform for large-scale production of integration-free human induced pluripotent stem cells in chemically defined culture. Sci Rep 5:11319
119. ThermoFischer Scientific (2017) Publications citing the Sendai Virus for iPSC Generation. http://www.thermofisher.com/us/en/home/life-science/stem-cell-research/induced-pluripo tent-stem-cells/sendai-virus-reprogramming/cytotune-publications.html?cid=fl-sendaipubs
120. Nishimura K et al (2011) Development of defective and persistent Sendai virus vector: a unique gene delivery/expression system ideal for cell reprogramming. J Biol Chem 286:4760–4771
121. Kato H et al (2006) Differential roles of MDA5 and RIG-I helicases in the recognition of RNA viruses. Nature 441:101–105
122. Kang X et al (2015) Effects of integrating and non-integrating reprogramming methods on copy number variation and genomic stability of human induced pluripotent stem cells. PLoS One 10:e0131128
123. Manzini S, Viiri LE, Marttila S, Aalto-Setala K (2015) A comparative view on easy to deploy non-integrating methods for patient-specific iPSC production. Stem Cell Rev 11:900–908
124. Choi J et al (2015) A comparison of genetically matched cell lines reveals the equivalence of human iPSCs and ESCs. Nat Biotechnol 33:1173–1181
125. Mallon BS et al (2014) Comparison of the molecular profiles of human embryonic and induced pluripotent stem cells of isogenic origin. Stem Cell Res 12:376–386
126. Guenther MG et al (2010) Chromatin structure and gene expression programs of human embryonic and induced pluripotent stem cells. Cell Stem Cell 7:249–257

127. Johannesson B et al (2014) Comparable frequencies of coding mutations and loss of imprinting in human pluripotent cells derived by nuclear transfer and defined factors. Cell Stem Cell 15:634–642
128. Salomonis N et al (2016) Integrated genomic analysis of diverse induced pluripotent stem cells from the progenitor cell biology consortium. Stem Cell Rep 7:110–125
129. Ma H et al (2014) Abnormalities in human pluripotent cells due to reprogramming mechanisms. Nature 511:177–183
130. Knoepfler P (2014) Stem cell pioneer Masayo Takahashi interview on iPS cells, clinical studies, & more. http://www.ipscell.com/2014/01/stem-cell-pioneer-masayo-takahashi-interview-on-ips-cells-clinical-studies-more/
131. Wang J et al (2015) Generation of clinical-grade human induced pluripotent stem cells in Xeno-free conditions. Stem Cell Res Ther 6:223
132. Wiley LA et al (2016) cGMP production of patient-specific iPSCs and photoreceptor precursor cells to treat retinal degenerative blindness. Sci Rep 6:30742
133. Marchetto MC et al (2009) Transcriptional signature and memory retention of human-induced pluripotent stem cells. PLoS One 4:e7076
134. Diecke S et al (2015) Novel codon-optimized mini-intronic plasmid for efficient, inexpensive, and xeno-free induction of pluripotency. Sci Rep 5:8081

Adv Biochem Eng Biotechnol (2018) 163: 23–38
DOI: 10.1007/10_2017_26
© Springer International Publishing AG 2017
Published online: 31 October 2017

Scalable Expansion of Pluripotent Stem Cells

Neta Lavon, Michal Zimerman, and Joseph Itskovitz-Eldor

Abstract Large-scale expansion of pluripotent stem cells (PSC) in a robust, well-defined, and monitored process is essential for production of cell-based therapeutic products. The transition from laboratory-scale protocols to industrial-scale production is one of the first milestones to be achieved in order to use both human embryonic stem cells (ESC) and induced pluripotent stem cells (iPSC) as the starting material for cellular products. The methods to be developed require adjustment of the culture platforms, optimization of culture parameters, and adaptation of downstream procedures. Optimization of expansion protocols and their scalability has become much easier with the design of bioreactor systems that enable continuous monitoring of culture parameters, continuous media change, and support software for automated control. This chapter highlights the common properties that are required for production of scalable, reproducible, homogeneous, and clinically suitable cell therapy products. We describe the available platforms for large-scale expansion of PSCs and parameters that should be considered when optimizing the expansion protocols in a scalable bioreactor. All the above are detailed in the light of the requirements and challenges of bringing a cell-based therapeutic product to the clinic and ultimately to the market. We discuss some considerations that should be taken into account, such as cost-effectiveness, good manufacturing practice, and regulatory guidelines.

N. Lavon (✉), M. Zimerman, and J. Itskovitz-Eldor
Kadimastem Ltd., Nes Ziona, Israel
e-mail: n.lavon@kadimastem.com; m.zimerman@kadimastem.com; itskovitz@gmail.com

Graphical Abstract

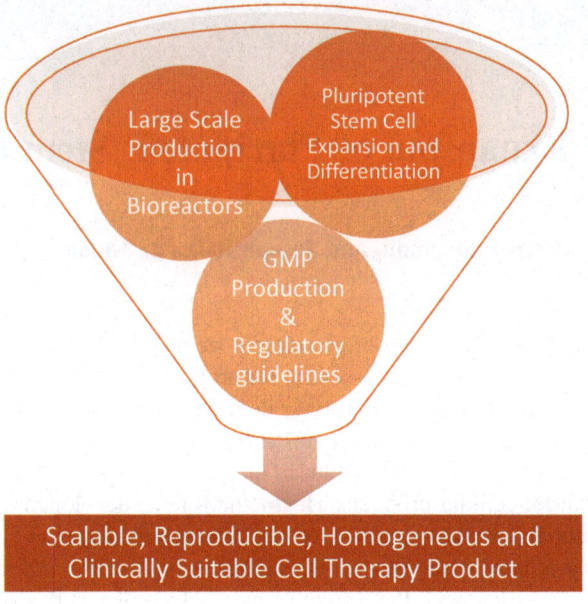

Keywords Bioreactor, Cell therapy, Embryonic stem cells, Large scale, Pluripotent stem cells, Process development, Production

Contents

1   Introduction

The dream of using human pluripotent cells (hPSCs) for cell therapy to treat human pathologies is close to becoming a reality. hPSCs, which include human embryonic stem cells (hESCs) and induced human pluripotent stem cells (hiPSCs), have the unique property of unlimited proliferation potential and the ability to differentiate to all cell types comprising the human body. Accordingly, hPSCs fit the requirement for cells that can serve as a source in cell therapy applications. Over the past 7 years, derivatives of hESCs have begun clinical trials for various pathologies such as spinal cord injury, myocardial infraction, and macular degeneration [1–3]. In 2015, a phase I/IIa clinical study in the USA showed long-term safety, graft survival, and possible biological activity of hESC-derived retinal pigmented epithelium (RPE) [2]. In Japan, hiPSC-derived RPE transplantation was performed but, unfortunately, the experiment was stopped because genomic aberrations were found in some iPSC lines [4]. A comprehensive review was published in 2016 of the current progress of hPSC clinical trials [5]. Overall, progress in the field of cell therapy using hPSCs gives a lot of hope and encouragement to keep developing various therapies.

Today, most of the therapies that have already advanced to the clinical stage require a fairly low number of cells per product. To use hPSCs as a source for therapies that either require a large number of cells per patient and/or aim at treating large numbers of patients, we need to develop industrial methods for controlled large-scale expansion of hPSCs. Developing such methodologies is a crucial step during development of therapies for conditions such as diabetes, pulmonary disease, and cardiomyopathy.

In this review, we discuss issues that are essential for the development of industrial large-scale expansion platforms of hPSCs for cell therapy. These include scalability, process development, growth methodologies, types of bioreactors and complimentary equipment, growth materials, and the compliance of all of the above with good manufacturing practice (GMP) and the requirements of regulatory authorities.

2   The History of Human Pluripotent Stem Cells

Embryonic stem cells from human origin (hESCs) were derived for the first time in 1998 from the inner cell mass of blastocyst-stage embryos [6]. The hESC lines had a substantial impact and advanced scientific research by supplying biological material for exploring the mechanisms of differentiation into the many cell types comprising the human body [7, 8]. Furthermore, hESCs have the potential to self-renew, which enables their expansion to an unlimited number of cells. Proof-of-concept studies demonstrating hESC cultivation in a dish provided the basis for developing reproducible protocols for differentiating hESCs into specific cell types, such as hepatocytes and oligodendrocytes [9, 10]. These cell types are now being used for drug development and drug toxicity studies, and as a source for cell therapy in pathological conditions [11, 12]. In 2007, a novel method for generating human pluripotent cells was described; introduction of a small set of transcription factors into a differentiated cell caused the somatic cell to revert to a pluripotent state [13]. When derived from a human source, these cells, termed "induced pluripotent stem cells" (iPSCs), were shown to have characteristics very similar to those of hESCs [14]. The hiPSC technology allows each person to be their own source of cells for autologous therapy. Furthermore, hiPSCs from patients' cells can serve as a unique source for studying disease mechanisms and exploring possible treatments [15].

Industrial development methodologies are necessary for the use of hESCs and hiPSCs as an unlimited source of cells for cell therapies. The following sections describe the sequential process of developing a cell therapy product, focusing on the large-scale expansion of pluripotent cells.

3   Requirements and Challenges in Cell Therapy Product Development

In the course of developing cell therapy products for large numbers of patients, there are several crucial steps to ensure the production of safe and effective product. The hPSC line has to have clinical potential, which requires full ethical approval for material collection, and documentation of the derivation and expansion procedures and quality control results [16]. Moreover, the entire production process requires the conversion of methods and materials to GMP compliance. Master cell bank (MCB) and working cell banks (WCBs) should be produced in a validated clean room. The WCB is the source for the production of large-scale reproducible batches of hPSCs to be released for the subsequent differentiation steps toward the final product. We review the production process from the industrial point of view, emphasizing the crucial step of large-scale expansion of hPSCs, which is essential for product development and its ultimate economic viability.

The differentiation protocol applied for directed differentiation into a desired lineage should be efficient and yield a high percentage of the required cell type [17, 18]. Some protocols require additional selection steps to enrich the culture with the desired cell type, either by collection of the required cells from the general population or by removal of nonrelevant cells [19]. The final product is released on the basis of its identity and potency. Elimination of undifferentiated pluripotent cells is essential in order to avoid any tumorigenic risks in the final product, which is an important general risk of hPSC-based therapies. The process of eliminating undifferentiated hPSCs to ensure the purity of the final product may be based on selection methods (e.g., using substances that selectively abolish pluripotent cells) or suicide gene technologies [20–23]. Generating the desired cell type is a great achievement in the process. However, additional steps are required to reach a final product suitable for initiating clinical trials. In some cases, successful transplantation and functional engraftment in the patient requires development of a scaffold or medical device to support the cells in vivo [24, 25]. Moreover, allogeneic transplants of hPSCs, in contrast to autologous transplants, require additional developmental steps to avoid immune rejection, such as wrapping the allogenic cells with an encapsulation device or putting patients on an immunosuppressive regime, as in current hESC clinical trials for diabetes and age-related macular degeneration [25, 26]. The stability of the final product is also a major concern. Frozen products are shipped fairly easily, but physicians need to be trained in the correct procedures for thawing and washing prior to transplantation. With fresh products, the stability over time dictates the ability to transport the product from the production facility to the clinical site. This time limit substantially affects the number of production sites and the cost of the final product.

At the end of this long journey of product development, the product's preclinical and clinical development starts. The product goes through safety studies in animals and three clinical phases in humans to prove its safety and efficacy prior to final approval for commercialization by regulatory authorities [27, 28]. A review published by the Process and Product Development Subcommittee of the International Society for Cellular Therapy nicely presents an introduction to the challenges of process development in cell-based therapies and a description of the tools available to address production issues [28].

4   Strategies for Scalable Expansion of Pluripotent Stem Cells

The diversity in potential cell sources (hESCs, hiPSCs, progenitor stem cells, adult stem cells, etc.), final therapeutic product (secreted molecules, single cells, aggregates, or encapsulated microtissues), culturing methodologies, and desired cell yield have led to the development of many culture platforms and extensive research on the optimization of culture conditions for each purpose.

The following basic features are generally needed for the production of a scalable, reproducible, homogeneous, and clinically suitable cell therapy product:

1. *Automatic online monitoring and control* enables tracking of culture conditions at all times and is essential for system optimization and better scalability. Moreover, an automated feedback looped system reduces the heterogeneity that can arise from manual systems that might suffer from human error.
2. *Scaling up* by moving to a larger vessel, instead of scaling out by use of multiple smaller vessels, minimizes the risk of contamination and aims at increasing homogeneity and robustness of the process. Ideally, a scalable system supports the transition from small scale to large scale without dramatic changes in culture conditions, thus enabling optimizations to be performed in small, more cost-efficient volumes.
3. *Dynamic culturing* allows improved mass transfer to ensure optimized circulation of oxygen and nutrients and homogenous distribution of cells. The mixing technique used should be optimized for the desired product. Current dynamic culture techniques use rotating platforms, air-based mixing, impeller-based stirring, and rocking. The main difference between these systems is the shear force applied on the cells.

4. *Continuous media change* is important in order to maintain the stability of the culture conditions, supply cells with fresh nutrients, and remove waste products. Sharp changes in growth factor concentration, oxygen levels, pH, and nutrients are limiting factors in the proliferation of hPSCs [29, 30]. Culture platforms that support continuous media change (perfusion) with the appropriate cell retention apparatus are essential for optimization of a large-scale system for cell therapy products.

5. *Cost effectiveness* of the process is affected by all the parameters detailed above, so the cost-to-yield-ratio should be evaluated for each system.

Optimized expansion processes should allow high-rate proliferation, keep the cells' pluripotency and genetic stability, and allow their differentiation to the cell type required as the final product. In the next subsections (Sect. 4.1–4.3), we review several large-scale expansion platforms for cell therapy applications, with emphasis on the properties listed above. We highlight the advantages and disadvantages of each platform and give some guidance on how to choose the one that best fits the desired product.

4.1   2D Static T-Flask-Based Culture

Traditionally, hPSCs were grown as a monolayer of cells in colonies, on either a fibroblast feeder layer or an extracellular matrix (ECM) that can support their growth and expansion. This method of T-flask-based culturing is scalable to some extent by increasing the surface area of the vessels and by scaling out and using multiple vessels. The 2D static T-flask-based system fits the requirements for products that require a small number of cells, such as autologous cell therapies, and can be used for example in personalized hiPSC-based products [31]. T-flask-based systems can be scaled out to create multilayered vessels for large-scale expansion for adherent cell culture. Multilayered 2D static vessels available on the market include Cell Factory™ (Thermoscientific), CellSTACK® (Corning), and Hyperflasks® (Corning). Controlled 2D systems are available on the market that allow automated monitoring and control of pH and oxygen levels and media circulation. These systems apply lower shear forces compared with stirred-tank bioreactors. The Integrity™ Xpansion® (ATMI) system is scalable up to 110,000 cm^2. Proof-of-concept experiments using the Xpansion® system for large-scale expansion of hESCs on feeder cells demonstrated the ability of the system to support culture expansion and maintain pluripotency [32].

In spite of these advances, static culture is limited compared with dynamic culture in scalability, online monitoring, and mass transfer, which limit the system's applicability for large-scale production processes.

4.2 Microcarrier-Based Dynamic Suspension Culture

Understanding the advantages of a dynamic suspension culture, and combining it with the advantages of a supporting matrix for hPSCs, led to the development of microcarriers (MCs) as a pseudosuspension culture platform. This technology has been extensively characterized for many cell types and different supporting matrices, as reviewed by Chen and colleagues [33]. When working with MCs, the shape, size, and surface properties of the MC have a significant effect on the expansion and differentiation ability of the cells [34, 35]. For example, Matrigel-coated, positively charged cylinder MCs and positively charged spherical MCs were found to support high cell yield and stable pluripotency of hESCs, but macroporous beads and small diameter (65 and 10 μm) spherical MCs did not [34].

Comparison between growing hPSCs in a 2D static system and an MC-based suspension system showed that MCs provide an advantage, resulting in a twofold higher expansion and a total yield of 3.5×10^6 cells/mL [36].

MC-based cultures are typically performed in bioreactor systems. These dynamic conditions allow better oxygen and nutrient transfer to the cells. On the other hand, the MC culture is not homogenous, hPSCs are not seeded equally on the surface of the MCs, and the aggregates tend to adhere to each other (our unpublished findings). The structures created because of the adherence of MCs result in heterogeneity of the culture and cause differential transfer of nutrients and oxygen, thus enhancing undesired differentiation processes. Even if the issue of heterogeneity can be solved, the most prominent disadvantage of MC-based platforms is the need to release the cells from the MCs prior to their transplantation or to find a solution that enables the use of MCs in the final product. The harvesting process for separating cells from MCs is expected to reduce the number and viability of the harvested cells. In terms of commercial viability of the product, implementation of an (potentially) unnecessary step to the production process complicates matters and raises process costs.

4.3 Aggregate-Based 3D Suspension Culture

In the late 1990s, it was first shown that dissociated hPSCs in nonadherent conditions are able to form 3D spheres called embryoid bodies, which support spontaneous differentiation of undifferentiated hESCs into the three established germ layers (ectoderm, mesoderm, and endoderm). Utilizing the ability of hPSCs to aggregate in static and dynamic platforms and with the discovery of the ROCK inhibitors' ability to increase the viability of single cells, researchers have shown that hPSCs can maintain their proliferative capacity and pluripotency over sequential passages in a matrix-free environment [37, 38]. Some different aggregate-based culture platforms are briefly described next.

3D suspension culture platforms based on orbital shaking are the easiest to establish and are useful for small-scale expansion and process development. In such systems, low-attachment tissue culture multiwell plates and Erlenmeyer flasks on orbital shakers are used to allow the testing of several growth conditions and cell concentrations in volumes of less than a milliliter to a few tens of milliliters. It was demonstrated that hiPSCs and hESCs cultured in six-well suspension plates placed on an orbital shaker increased sixfold in number within 4 days. With the use of a defined media, they maintained their pluripotency for up to 17 passages [39]. However, these systems are not monitored and do not permit continuous media change, but the transition from shaker-based platforms to other 3D aggregate-based bioreactors is a common practice.

Spinner flask vessels are useful in scaling the process up to hundreds of milliliters. Here, dynamic conditions are achieved using rotating impellers similar to those in stirred-tank bioreactors. They are not automatically monitored and controlled; hence, reproducibility is hard to reach.

Automatic real-time monitoring and control of culture conditions (e.g., oxygen concertation, pH, temperature, nutrient concentration, mixing speed, and impeller design) can facilitate the development of an efficient, homogeneous, and reproducible culture process. Thus, bioreactors equipped with various sensors and controlling software are the next step in the scalable expansion of hPSCs.

Rotary cell culture systems such as CELLON (Synthecon) are rotating 3D vessels. Such systems provide a dynamic, low-shear stressed environment, controlled oxygenation, continuous media change, and good mass transfer to support PSC expansion and differentiation [40–43]. However, the size of the system is limited (up to 150 mL) and scale-up is not easy.

Cell culture bags are another option for the large-scale expansion of mammalian cells. Gently rocking the cells back and forth produces homogenous aggregation. Such systems have been developed over the years, from the Wave Bioreactor (GE Healthcare), to the BIOSTAT CultiBags (Sartorious) and the AppliFlex (Applikon). Current culture bag systems enable control of temperature, gassing, and pH. Scaling-up is easy and, most importantly, cells are subject to relatively low shear forces. The system has gained popularity and is suitable for applications such as murine iPSC differentiation to cardiomyocytes in suspension culture and mesenchymal stem cell expansion on MCs [44, 45]. The Wave Bioreactor is a disposable alternative for large-scale expansion of hPSCs. The single-use vessels do not require between-batch validation or cleaning.

Stirred tank bioreactors are still considered the best platform for 3D culture of hPSCs. They are simple to scale up and are fully controlled, enabling cell-specific optimization of culture parameters and commercial viability of large-scale production of cell therapy products. Stirred-tank bioreactors have a basic design consisting of a glass and/or a single-use plastic vessel with a designated head plate; rotating impeller; probes for monitoring temperature, pH, and dissolved oxygen; and a gassing system. Such systems are very robust and able to achieve reproducible results once optimized for their initial cell density, culture media, culture conditions, and stirring technique [39, 46]. Once the right parameters for cultivation in small-scale bioreactors are found, scale-up of the method should be quite straightforward and allow a volume increase up to several hundred liters. Parameters such as the impeller design, sensor/probe height inside the vessel, gassing technique, and perfusion rate may require adjustment when moving to larger volumes to enable the homogenous distribution of cells and accurate online monitoring of culture parameters.

In reviewing the crucial culture parameters, we focus on the stirred-tank bioreactor, but it is important to note that optimization of culture conditions such as temperature, pH, and oxygen as well as vessel design and mechanical forces should be considered and modified in each selected platform. A comparison of different 2D static and 3D suspension culturing systems is shown in Fig. 1.

A	Automatic online monitor and control	Scalable	Dynamic culturing	Continuous media change
2D static T-flasks based culture (B)	To some extent	To some extent	NO	In some cases
Orbital shaker based suspension culture platforms (C)	NO	NO	YES	NO
Spinner vessels (D)	NO	YES	YES	NO
Rotary cell culture (RCC) systems (E)	YES	To some extent	YES	YES
Cell culture bags (F)	YES	YES	YES	YES
Stirred tank bioreactors (G)	YES	YES	YES	YES

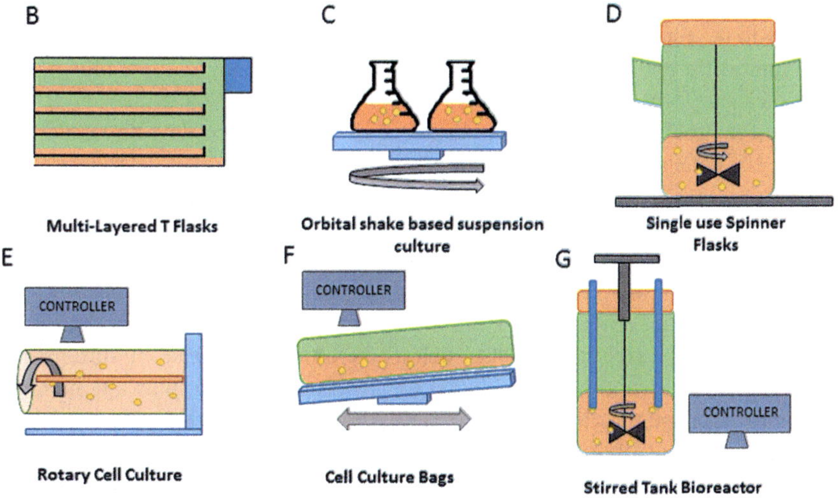

B — Multi-Layered T Flasks
C — Orbital shake based suspension culture
D — Single use Spinner Flasks
E — Rotary Cell Culture
F — Cell Culture Bags
G — Stirred Tank Bioreactor

Fig. 1 Platforms for large-scale expansion of hPSCs. (**a**) Comparison of the various platforms and their properties. (**b**) 2D static T-flask-based cultures are scaled out, adding more layers and enlarging the culturing surface. In the Xpansion® (ATMI) system, continuous media change and online monitoring is available. (**c**) 3D-aggregate culturing on orbital shakers is suitable for process development and small-scale protocol optimization. (**d**) Spinner flasks can act as a preliminary step before moving to a stirred-tank bioreactor because the agitation techniques are similar. However, the system is not monitored and does not allow continuous media change. (**e**) Rotary cell culture system is suitable for small-scale optimization of hPSC expansion and differentiation. (**f**) Cell culture bags are applicable for GMP manufacturing. They are single-use and equipped with online sensors and a perfusion system. (**g**) Stirred-tank bioreactors are widely used and easily scaled up. They are extremely flexible and suitable for multiple applications, enabling continuous media change and online control and monitoring of culture parameters

5 Optimization of Culture Conditions

In this section, we highlight some parameters (shear stress, aggregate diameter and homogeneity, oxygen concentration, and pH) found to be important for the optimization of hPSC cultivation in stirred-tank bioreactors. These parameters should therefore be considered when optimizing systems for the large-scale production of pluripotent cells.

Shear stress is a major concern when culturing hPSCs in stirred-tank bioreactors because the cells are very sensitive to hydrodynamic forces. Shear stress can increase cell death, decrease the ability of cells to aggregate, and induce differentiation instead of allowing maintenance of the pluripotency state [47]. Shear stress is influenced by impeller design, impeller diameter (shear stress increases with increase in diameter), impeller placement (height) inside the vessel, agitation speed, presence of gas bubbles, and the presence of probes that create mechanical obstacles in the reactor vessel [48–50]. Cell damage occurs when the turbulent Kolmogorov eddy size is equal to the diameter of cell aggregates; this parameter decreases as the agitation speed increases, thus making aggregates more sensitive to lower agitation speeds than single cells cultured in suspension. Moreover, the optimal hydrodynamic shear stress differs in different cell types and should be adapted according to the final product. That is, transition of hPSCs (at the pluripotent state) toward their differentiation into lineage-specific progenies may require adjustment of applied shear forces [51–53].

The current goal of large-scale expansion of hPSCs in stirred-tank bioreactors is to obtain 10^7 cells/mL. Achieving high process efficiency requires relatively high cell densities in a given process scale. This must be accompanied by adequate transfer of oxygen. In small process volumes of up to few liters, gassing through the overhead space maybe sufficient. In large vessels, a submerged gassing strategy is required, which can increase shear stress as a result of gas bubbles, known to be detrimental to neighboring cells because of bubble-burst. Moreover, submerged gassing into a media rich in proteins (such as the established hPSC culture media Essential 8, mTESR, and other products) can result in foaming, thus augmenting bubble-burst damage to cells. The use of antifoaming agents such as Pluronic F68, Antifoam C, methylcellulose, or polyethylene glycol protects cells from foaming damage and should be tested in the hPSC culture system [49, 54, 55]. Note that antifoaming products could cause cell toxicity at certain concentrations and should be evaluated before use.

Aggregate diameter and homogeneity are crucial in the establishment of a robust industrial-scale bioprocess. Aggregates exceeding the 300 µm diameter threshold experience hypoxia and low nutrient/growth-factor concentration in the core of the aggregate, reducing viability and pluripotency [56]. Heterogeneous aggregate sizes might lead to heterogeneous conditions as a result of concentration-dependent cues to cells in different locations within the aggregates. These different cues are suspected to result in heterogeneous populations of cells with varied levels of pluripotency and might affect their lineage-specific differentiation Aggregate diameter can be monitored via offline sampling and microscopy-based analysis. Aggregate size is controlled by inoculum cell density and agitation speed, with lower agitation speeds resulting in bigger aggregates. Homogeneity and control of aggregate size can also be controlled by impeller design [57]. Impellers are found in various shapes and sizes, but the most common in stem cell applications are marine and pitched impellers. The number of blades and the angle in which they are positioned determines the distribution of aggregates inside the dynamic culture, size of the aggregates, expansion capacity, and amount of shear force applied on the cells.

Oxygen concentration can affect the proliferation, chromosomal damage, and differentiation of hPSCs. Some papers suggest that "hypoxic" conditions (3–5% O_2) are advantageous for PSC proliferation in comparison to 21% O_2, but others show no effect [58–62, 63].

pH levels also influence cell proliferation and pluripotency. The accumulation of lactate and the reduction in pH is harmful to cells and can decrease cell viability. In murine ESCs, a lower pH of 6.8 was better for preserving cell pluripotency but significantly damaged cardiac differentiation compared with a slightly higher pH of 7.1 [64]. Using the automated control system of the stirred-tank bioreactor, pH levels could be corrected in real time. Adjustment of pH can be achieved by CO_2 gassing into the media, titration of a strong base (e.g., NaOH) or bicarbonate, continuous perfusion that removes waste products from the culture vessel and in turn adjusts the pH, or a combination of several of the above strategies.

All the above parameters should be optimized for the expansion of pluripotent cells and probably require adjustment for the large-scale differentiation process to be followed. For example, an increase in oxygen concentration is beneficial for the final maturation of beta-like insulin secreting cells [65]; therefore, elevating the oxygen levels in the bioreactor in the last stages of differentiation can improve the functionality of the final product.

6   Development of Cell-Based Therapies: Considerations and Future Aspects

Cell-based therapies require large quantities of PSCs. A recent publication shows that 3D matrix-free suspension cultures of hPSC aggregates can be substantially expanded in a 100-mL stirred-tank bioreactor [29]. Improving and expanding the monitoring and control systems is expected to enhance culture stability and allow higher cell yields.

Extensive work has been done to enable the conversion of protocols from laboratory scale to the large industrial scale. For that goal, system optimization is required for bioreactor design to support cell culture homogeneity, transition to single-use vessels, perfusion systems for continuous media change, and integration of online monitoring sensors. All of these have been developed and are still being improved to facilitate the transition to GMP large-scale robust cell production. However, to develop a fully automated large-scale closed system that will give reproducible results, monitoring of key metabolites in real time is necessary. In-vessel measurement of glucose consumption and lactate production together with cell density, oxygen consumption, and acidification rate all help in maintaining culture stability during large-scale expansion and in reaching the goal of product release much quicker. Online monitoring of cell density and viability, for example, are very important in increasing the reproducibility of the production process by preventing the human-based variability associated with manual sampling and aggregate dissociation offline. We expect these improvements in the optimization and reproducibility of the production process and the final product to assist in the process of product approval by the regulatory authorities.

Moreover, the ability of the culture platform to support both the expansion and differentiation of hPSCs is very important. Bioreactor flexibility refers to the ability of the system to be adjusted according to the required culture parameters. The ability to change agitation/mixing speed, gassing technique, gas composition, probe height, and growth factor concentration easily in the same vessel is very important for accurate adaption of a bioreactor system to the complex culture needs of hPSCs and their respective progenies. Changes in culture parameters may be required in later differentiation steps, thus system flexibility is important and even necessary to avoid the need to move cells from one culture system to another. Such a move can complicate the production process and increase production costs.

During development of expansion and differentiation protocols at the typical laboratory scale, researchers need to keep in mind that an easier transition to clinical production could be achieved if they use GMP-compliant materials at an early stage. There should always be a preference for small molecules over (recombinant) proteins, as chemical compounds are typically more stable and cheaper to produce. If proteins cannot be replaced, a synthetic or a recombinant variant of the molecule is preferable. Ready-to-use, chemically defined, and xeno-free media are more easily approved by regulatory authorities and require less testing and supporting documentation. The use of GMP-compliant materials together with single-use fully automated bioreactors that are optimized to maintain culture stability, pluripotency, and final differentiation should all be aspired to.

References

1. Priest CA, Manley NC, Denham J, Wirth III ED, Lebkowski JS (2015) Preclinical safety of human embryonic stem cell-derived oligodendrocyte progenitors supporting clinical trials in spinal cord injury. Regen Med 10(8):939–958
2. Schwartz SD et al. (2015) Human embryonic stem cell-derived retinal pigment epithelium in patients with age-related macular degeneration and Stargardt's macular dystrophy: follow-up of two open-label phase 1/2 studies. Lancet 385(9967):509–516
3. Menasché P, Vanneaux V, Hagège A, Bel A, Cholley B, Cacciapuoti I, Parouchev A, Benhamouda N, Tachdjian G, Tosca L, Trouvin JH, Fabreguettes JR, Bellamy V,

Guillemain R, Suberbielle Boissel C, Tartour E, Desnos M, Larghero J (2015) Human embryonic stem cell-derived cardiac progenitors for severe heart failure treatment: first clinical case report. Eur Heart J 36(30):2011–2017

4. KYODO (2015) First iPS cell transplant patient makes progress one year on
5. Trounson A, DeWitt ND (2016) Pluripotent stem cells progressing to the clinic. Nat Rev Mol Cell Biol 17(3):194–200
6. Thomson JA et al. (1998) Embryonic stem cell lines derived from human blastocysts. Science 282(5391):1145–1147
7. Itskovitz-Eldor J, Schuldiner M, Karsenti D, Eden A, Yanuka O, Amit M, Soreq H, Benvenisty N (2000) Differentiation of human embryonic stem cells into embryoid bodies compromising the three embryonic germ layers. Mol Med 6(2):88–95
8. Schuldiner M et al. (2000) Effects of eight growth factors on the differentiation of cells derived from human embryonic stem cells. Proc Natl Acad Sci U S A 97(21):11307–11312
9. Lavon N, Yanuka O, Benvenisty N (2004) Differentiation and isolation of hepatic-like cells from human embryonic stem cells. Differentiation 72(5):230–238
10. Izrael M, Zhang P, Kaufman R, Shinder V, Ella R, Amit M, Itskovitz-Eldor J, Chebath J, Revel M (2007) Human oligodendrocytes derived from embryonic stem cells: effect of noggin on phenotypic differentiation in vitro and on myelination in vivo. Mol Cell Neurosci 34 (3):310–323
11. Chris Mason DAB, Culme-Seymour EJ, Davie NL (2011) Cell therapy industry: billion dollar global business with unlimited potential. Regen Med 6(3):265–272
12. Davidson MD, Ware BR, Khetani SR (2015) Stem cell-derived liver cells for drug testing and disease modeling. Discov Med 19(106):349–358
13. Takahashi K, Yamanaka S (2006) Induction of pluripotent stem cells from mouse embryonic and adult fibroblast cultures by defined factors. Cell 126(4):663–676
14. Buganim Y, Faddah DA, Jaenisch R (2013) Mechanisms and models of somatic cell reprogramming. Nat Rev Genet 14(6):427–439
15. Bellin M, Marchetto CM, Gage FH, Mummery CL (2012) Induced pluripotent stem cells: the new patient? Nat Rev Mol Cell Biol 13:713–726
16. ISSCR (2016) Guidlines for stem cell research and clinical translation. International Society for Stem Cell Research, Skokie. Available at http://www.isscr.org/professional-resources/policy/2016-guidelines/guidelines-for-stem-cell-research-and-clinical-translation
17. Kempf H, Olmer R, Kropp C, Rückert M, Jara-Avaca M, Robles-Diaz D, Franke A, Elliott DA, Wojciechowski D, Fischer M, Lara AR, Kensah G, Gruh I, Haverich A, Martin U, Zweigerdt R (2014) Controlling expansion and cardiomyogenic differentiation of human pluripotent stem cells in scalable suspension culture. Stem Cell Rep 3(6):1132–1146
18. Kempf H, Andree B, Zweigerdt R (2016) Large-scale production of human pluripotent stem cellderived cardiomyocytes. Adv Drug Deliv Rev 15:18–30
19. Kelly OG et al. (2011) Cell-surface markers for the isolation of pancreatic cell types derived from human embryonic stem cells. Nat Biotechnol 29(8):750–756
20. Schuldiner M, Itskovitz-Eldor J, Benvenisty N (2003) Selective ablation of human embryonic stem cells expressing a "suicide" gene. Stem Cells 21(3):257–265
21. Kropp EM, Oleson B, Broniowska KA, Bhattacharya S, Chadwick AC, Diers AR, Hu Q, Sahoo D, Hogg N, Boheler KR, Corbett JA, Gundry RL (2015) Inhibition of an NAD+ salvage pathway provides efficient and selective toxicity to human pluripotent stem cells. Stem Cells Transl Med 4(5):483–493
22. Ben-David U et al. (2013) Selective elimination of human pluripotent stem cells by an oleate synthesis inhibitor discovered in a high-throughput screen. Cell Stem Cell 12(2):167–179
23. Boheler KR, Bhattacharya S, Kropp EM, Chuppa S, Riordon DR, Bausch-Fluck D, Burridge PW, Wu JC, Wersto RP, Chan GC, Rao S, Wollscheid B, Gundry RL (2014) A human pluripotent stem cell surface N-glycoproteome resource reveals markers, extracellular epitopes, and drug targets. Stem Cell Rep 3(1):185–203

24. Mitragotri S, Burke PA, Langer R (2014) Overcoming the challenges in administering biopharmaceuticals: formulation and delivery strategies. Nat Rev Drug Discov 13(9):655–672
25. Agulnick AD, Ambruzs D, Moorman MA, Bhoumik A, Cesario RM, Payne JK, Kelly JR, Haakmeester C, Srijemac R, Wilson AZ, Kerr J, Frazier MA, Kroon EJ, D'Amour KA (2015) Insulin-producing endocrine cells differentiated in vitro from human embryonic stem cells function in macroencapsulation devices in vivo. Stem Cells Transl Med 4(10):1214–1222
26. Schwartz SD et al. (2016) Subretinal transplantation of embryonic stem cell-derived retinal pigment epithelium for the treatment of macular degeneration: an assessment at 4 years. Investig Ophthalmol Vis Sci 57(5):ORSFc1–ORSFc9
27. Salmikangas P et al. (2015) Marketing regulatory oversight of advanced therapy medicinal products (ATMPs) in Europe: the EMA/CAT perspective. Adv Exp Med Biol 871:103–130
28. Campbell A et al. (2015) Concise review: process development considerations for cell therapy. Stem Cells Transl Med 4(10):1155–1163
29. Kropp C, Kempf H, Halloin C, Robles-Diaz D, Franke A, Scheper T, Kinast K, Knorpp T, Joos TO, Haverich A, Martin U, Zweigerdt R, Olmer R (2016) Impact of feeding strategies on the scalable expansion of human pluripotent stem cells in single-use stirred tank bioreactors. Stem Cells Transl Med 5(10):1289–1301
30. Yeo D, Kiparissides A, Cha JM, Aguilar-Gallardo C, Polak JM, Tsiridis E, Pistikopoulos EN, Mantalaris A (2013) Improving embryonic stem cell expansion through the combination of perfusion and bioprocess model design. Plos One 8(12):e81728
31. Kamao H et al. (2014) Characterization of human induced pluripotent stem cell-derived retinal pigment epithelium cell sheets aiming for clinical application. Stem Cell Rep 2(2):205–218
32. Roberts I, Moens N, Moncaubeig F, Egloff M, Coffey P, Mason C. The importance of using small scale bioreactor mimics to scale up human embryonic stem cell culture. Available at http://studylib.net/doc/10488641/the-importance-of-using-small-scale-bioreactor-mimics-to-...
33. Chen AK, Reuveny S, Oh SK (2013) Application of human mesenchymal and pluripotent stem cell microcarrier cultures in cellular therapy: achievements and future direction. Biotechnol Adv 31(7):1032–1046
34. Chen AK et al. (2011) Critical microcarrier properties affecting the expansion of undifferentiated human embryonic stem cells. Stem Cell Res 7(2):97–111
35. Sart S et al. (2013) Modulation of mesenchymal stem cell actin organization on conventional microcarriers for proliferation and differentiation in stirred bioreactors. J Tissue Eng Regen Med 7(7):537–551
36. Oh SK et al. (2009) Long-term microcarrier suspension cultures of human embryonic stem cells. Stem Cell Res 2(3):219–230
37. Amit M, Chebath J, Margulets V, Laevsky I, Miropolsky Y, Shariki K, Peri M, Blais I, Slutsky G, Revel M, Itskovitz-Eldor J (2010) Suspension culture of undifferentiated human embryonic and induced pluripotent stem cells. Stem Cell Rev 6(2):248–259
38. Steiner D, Khaner H, Cohen M, Even-Ram S, Gil Y, Itsykson P, Turetsky T, Idelson M, Aizenman E, Ram R, Berman-Zaken Y, Reubinoff B (2010) Derivation, propagation and controlled differentiation of human embryonic stem cells in suspension. Nat Biotechnol 28 (4):361–364
39. Olmer R et al. (2010) Long term expansion of undifferentiated human iPS and ES cells in suspension culture using a defined medium. Stem Cell Res 5(1):51–64
40. Come J, Nissan X, Aubry L, Tournois J, et al. (2008) Improvement of culture conditions of human embryoid bodies using a controlled perfused and dialyzed bioreactor system. Tissue Eng Part C Methods 14(4):289–298
41. Fridley KM et al. (2010) Unique differentiation profile of mouse embryonic stem cells in rotary and stirred tank bioreactors. Tissue Eng Part A 16(11):3285–3298
42. Gerecht-Nir S, Cohen S, Itskovitz-Eldor J (2004) Bioreactor cultivation enhances the efficiency of human embryoid body (hEB) formation and differentiation. Biotechnol Bioeng 86(5):493–502

43. Wang X et al. (2006) Scalable producing embryoid bodies by rotary cell culture system and constructing engineered cardiac tissue with ES-derived cardiomyocytes in vitro. Biotechnol Prog 22(3):811–818
44. Timmins NE et al. (2012) Closed system isolation and scalable expansion of human placental mesenchymal stem cells. Biotechnol Bioeng 109(7):1817–1826
45. Correia C et al. (2014) Combining hypoxia and bioreactor hydrodynamics boosts induced pluripotent stem cell differentiation towards cardiomyocytes. Stem Cell Rev 10(6):786–801
46. Singh H et al. (2010) Up-scaling single cell-inoculated suspension culture of human embryonic stem cells. Stem Cell Res 4(3):165–179
47. Wolfe RP et al. (2012) Effects of shear stress on germ lineage specification of embryonic stem cells. Integr Biol 4(10):1263–1273
48. Gilbertson JA et al. (2006) Scaled-up production of mammalian neural precursor cell aggregates in computer-controlled suspension bioreactors. Biotechnol Bioeng 94(4):783–792
49. Schroeder M et al. (2005) Differentiation and lineage selection of mouse embryonic stem cells in a stirred bench scale bioreactor with automated process control. Biotechnol Bioeng 92(7):920–933
50. Fridley KM, Kinney MA, McDevitt TC (2012) Hydrodynamic modulation of pluripotent stem cells. Stem Cell Res 3(6):45
51. Bauwens C et al. (2005) Development of a perfusion fed bioreactor for embryonic stem cell-derived cardiomyocyte generation: oxygen-mediated enhancement of cardiomyocyte output. Biotechnol Bioeng 90(4):452–461
52. Cormier JT, Nieden N, Rancourt DE, Kallos MS (2006) Expansion of undifferentiated murine embryonic stem cells as aggregates in suspension culture bioreactors. Tissue Eng 12(11):3233–3245
53. Youn BS et al. (2005) Large-scale expansion of mammary epithelial stem cell aggregates in suspension bioreactors. Biotechnol Prog 21(3):984–993
54. Wu J (1995) Mechanisms of animal cell damage associated with gas bubbles and cell protection by medium additives. J Biotechnol 43(2):81–94
55. Thomas CR, Zhang Z, Al-Rubeai M (1992) Effect of Pluronic F-68 on the mechanical properties of mammalian cells. Enzym Microb Technol 14(12):980–983
56. Wu J, Rostami MR, Cadavid Olaya DP, Tzanakakis ES (2014) Oxygen transport and stem cell aggregation in stirred-suspension bioreactor cultures. PLoS One 9(7):e102486
57. Olmer R et al. (2012) Suspension culture of human pluripotent stem cells in controlled, stirred bioreactors. Tissue Eng Part C Methods 18(10):772–784
58. Abaci HE et al. (2010) Adaptation to oxygen deprivation in cultures of human pluripotent stem cells, endothelial progenitor cells, and umbilical vein endothelial cells. Am J Physiol Cell Physiol 298(6):C1527–C1537
59. Chen HF, Kuo H, Chen W, Wu FC, Yang YS, Ho HN (2009) A reduced oxygen tension (5%) is not beneficial for maintaining human embryonic stem cells in the undifferentiated state with short splitting intervals. Hum Reprod 24(1):71–80
60. Ezashi T, Das P, Roberts RM (2005) Low O2 tensions and the prevention of differentiation of hES cells. Proc Natl Acad Sci U S A 102(13):4783–4788
61. Mohyeldin A, Garzon-Muvdi T, Quinones-Hinojosa A (2010) Oxygen in stem cell biology: a critical component of the stem cell niche. Cell Stem Cell 7(2):150–161
62. Shyh-Chang N, Daley GQ, Cantley LC (2013) Stem cell metabolism in tissue development and aging. Development 140(12):2535–2547
63. Simon MC, Keith B (2008) The role of oxygen availability in embryonic development and stem cell function. Nat Rev Mol Cell Biol 9(4):285–296
64. Teo A, Mantalaris A, Lim M Influence of culture pH on proliferation and cardiac differentiation of murine embryonic stem cells. Biochem Eng J 90:8–15
65. Cechin S et al. (2014) Influence of in vitro and in vivo oxygen modulation on beta cell differentiation from human embryonic stem cells. Stem Cells Transl Med 3(3):277–289

Adv Biochem Eng Biotechnol (2018) 163: 39–70
DOI: 10.1007/10_2017_30
© Springer International Publishing AG 2017
Published online: 26 October 2017

Scalable Cardiac Differentiation of Pluripotent Stem Cells Using Specific Growth Factors and Small Molecules

Henning Kempf and Robert Zweigerdt

Abstract The envisioned routine application of human pluripotent stem cell (hPSC)-derived cardiomyocytes (CMs) for therapies and industry-compliant screening approaches will require efficient and highly reproducible processes for the mass production of well-characterized CM batches.

On their way toward beating CMs, hPSCs initially undergo an epithelial-to-mesenchymal transition into a primitive-streak (PS)-like population that later gives rise to all endodermal and mesodermal lineages, including cardiovascular progenies (CVPs). CVPs are multipotent and possess the capability to give rise to all major cell types of the heart, including CMs, endothelial cells, cardiac fibroblasts, and smooth muscle cells. This article provides an historical overview and describes the stepwise development of protocols that typically result in the appearance of beating CMs within 7–12 days of hPSC differentiation.

We describe the development of directed and closely controlled cardiomyogenic differentiation, which now enables the induction of >90% CM purity without further lineage enrichment. Although secreted lineage specifiers (revealed from developmental biology) were initially used, we outline the advantages of chemical pathway modulators, as defined by more recent screening approaches. Subsequently, we discuss the use of defined culture media for upscaling the production of hPSC-CMs in controlled bioreactors and how this, in principle, unlimited source of human CMs can be used to progress heart regeneration and stimulate the drug discovery pipeline.

H. Kempf (✉) and R. Zweigerdt
Leibniz Research Laboratories for Biotechnology and Artificial Organs (LEBAO), Hannover Medical School (MHH), Carl-Neuberg-Str. 1, 30625 Hannover, Germany
e-mail: Kempf.henning@mh-hannover.de; zweigerdt.robert@mh-hannover.de

Graphical Abstract

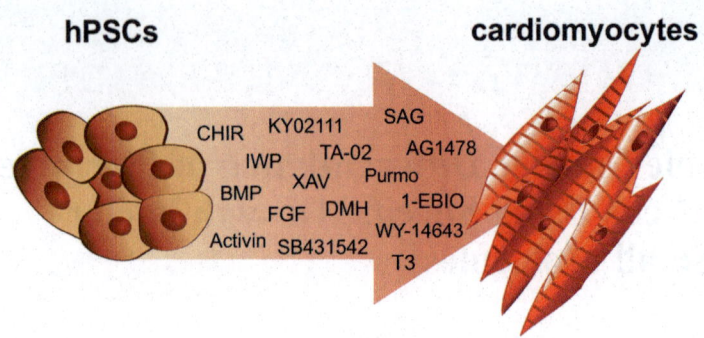

Keywords Bioreactor, Cardiac differentiation, Cardiomyocyte, hPSC, hPSC-CM, Induced pluripotent stem cells, Mass production, Primitive streak, Upscaling

Contents

Abbreviations

1-EBIO	1-Ethyl-2-benzimidazolone
bFGF	Basic fibroblast growth factor (FGF2)
BMP	Bone morphogenic protein
BSA	Bovine serum albumin
bSF	Basal serum-free
CDM	Chemically defined medium
CDX1/2	Caudal-type homeobox 1/2
CER1	Cerberus
CHIR	CHIR99021
CK1	Caseine kinase 1
CM	Cardiomyocyte
cTNT	Cardiac troponin T
CVP	Cardiovascular progenitor
DKK1	Dickkopf 1

EGFR	Epidermal growth factor receptor
EMT	Epithelial-to-mesenchymal
EOMES	Eomesodermin
Fz	Frizzled
GSK3	Glycogen synthase kinase 3
hESC	Human embryonic stem cell
hiPSC	Human induced pluripotent stem cell
hPSC	Human pluripotent stem cell
ICAT	Inhibitor of β-catenin and TCF-4
IDE	Inducer of definitive endoderm
IGF	Insulin growth factor
ISL1	Islet1
IWP2	Inhibitor of WNT production 1
IWR1	Inhibitor of WNT response 1
LEFTY1	Left-right determination factor 1
LRP5/6	Low density lipoprotein receptor-related protein co-receptor 5/6
MESP1	Mesoderm posterior 1 homolog
MHC	Myosin heavy chain
MIXL1	Mix paired-like homeobox 1
MLC2v	Myosin light chain 2v
MSX1/2	msh homeobox 1
NCAM	Neural cell adhesion molecule
NKX2.5	NK2 homeobox 5
OCT3/4	Octamer binding transcription factor
PORC	Porcupine
PS	Primitive streak
ROCK	Rho-associated kinase
SCF	Stem cell factor
SIRPα	Signal-regulatory protein alpha
T	T-brachyury
T3	Tri-iodo-L-thyronine
TBX5	T-box transcription factor
TGF	Transforming growth factor
TNKS	Tankyrase
VCAM1	Vascular cell adhesion molecule 1
VEGF	Vascular endothelial growth factor
WNT	Wingless protein
WRE	WNT response element

1 Introduction

Differentiation of human pluripotent stem cells (hPSCs; collectively referring to human embryonic and induced pluripotent stem cells; hESC and hiPSC) in vitro recapitulates key aspects of early human development. On their way toward beating cardiomyocytes, hPSCs initially undergo an epithelial-to-mesenchymal (EMT) transition into a primitive-streak (PS)-like population that later gives rise to all endodermal and mesodermal lineages. The PS stage is marked by the expression of typical PS markers such as MIXL1, T-brachyury, and EOMES within 1–2 days after induction of differentiation, depending on culture conditions. Recent studies provide evidence that this early PS population is already primed and patterned into distinct sublineages (Fig. 1) that range from endoderm and lateral mesoderm to late populations such as the presomitic mesoderm [1–3]. The PS-like patterning of hPSCs in vitro thus represents a model of both spatially and temporally distinct populations of the in vivo human PS (Fig. 1). Proposed markers characterizing early PS subsets include NANOG, SOX2, CDX1/2, and MSX1/2 [3–5].

Following the emergence of the (lateral) mesoderm from the PS, further specification into cardiac mesoderm and cardiac progenitors can be induced by appropriate culture conditions (Figs. 1 and 2). These cardiovascular progenitors (CVPs) are multipotent and possess the capability to give rise to the major cell types of the heart, including cardiomyocytes (CMs), endothelial cells, cardiac fibroblasts, and smooth muscle cells. Regarding CMs, beating cells are typically obtained after 7–12 days of differentiation, depending on the differentiation protocol. However, these early CMs exhibit a rather immature phenotype with respect to morphology, sarcomere organization, gene expression pattern, metabolic activity, and resulting electrophysiological and contractile characteristics.

This article describes how recent research has enabled direct and close control of cardiomyogenic differentiation of hPSCs by utilizing either secreted lineage

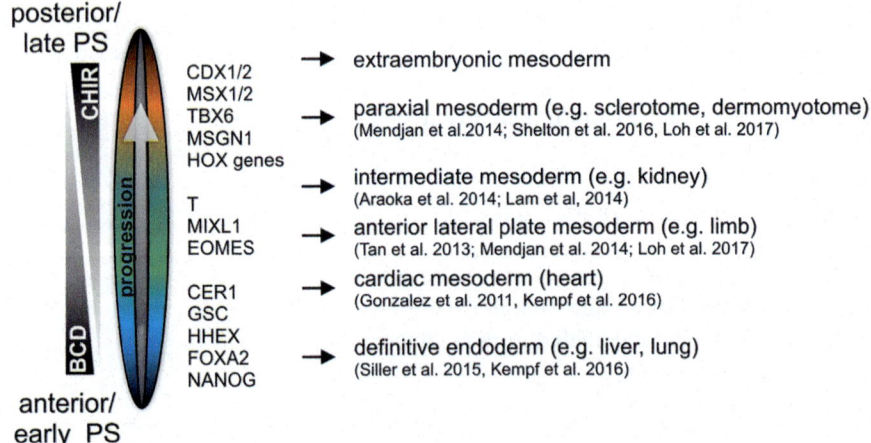

Fig. 1 Primitive streak (PS)-like patterning of hPSCs in vitro. PS patterning along the anterior–posterior axis can be induced by WNT pathway activation using, for example, CHIR99021 (*CHIR*) in combination with controlled bulk cell density (*BCD*). Notably, PS patterning of hPSCs can be observed readily within 24 h at the gene expression level [1]. Typical marker genes characterizing the PS stage and prospective lineages are indicated along the PS axis

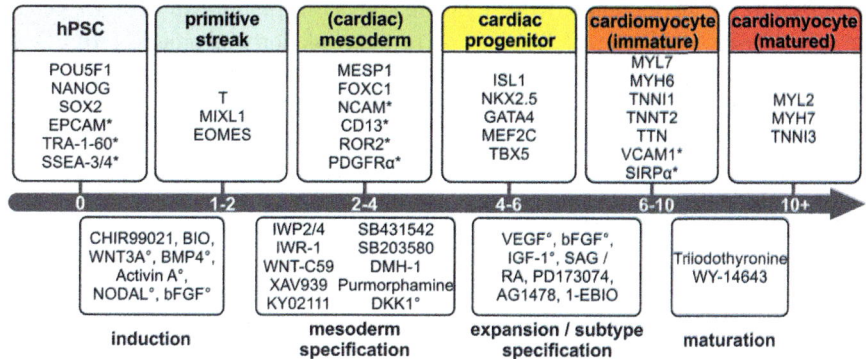

Fig. 2 Time course of a typical cardiac differentiation protocol for hPSCs. The *upper boxes* indicate the various stages and respective biomarkers from pluripotency (*left*) toward CMs (*right*). *Lower boxes* indicate key molecules for directing this process in vitro. Surface markers are marked by an *asterisk* and growth factors by a *circle*

specifiers revealed from developmental biology and/or the use of chemical pathway modulators identified through screening approaches. These findings are organized chronologically from the pluripotent stage of hPSCs toward their differentiation into functional CMs and subsequent attempts to progress their maturation, as graphically summarized in Fig. 2. Moreover, we highlight the practical use of new findings for the development of processes for upscaling the production of hPSC-derived cardiomyocytes (hPSC-CMs) in controlled bioreactors. Prominent examples are given to demonstrate how this novel, formerly unavailable source of human CMs can now be used to progress heart regeneration and stimulate the drug discovery pipeline.

2 Current Technologies, Standards, and Strategies

2.1 Recombinant Growth Factors to Direct Cardiac Cell Fate

Most established cardiac differentiation protocols implement findings from developmental biology regarding mimicking specific aspects of the complexity of developmental processes in vitro. The first directed protocols were based on the sequential addition of growth factors regulating key signaling pathways discovered in animal (flies, frogs, fish, and mice) models of cardiac development. Among these growth factors were secreted agonists belonging to the transforming growth factor (TGF) superfamily of growth factors, including bone morphogenic protein (BMP)2/4, Activin A [6, 7], Nodal [5], basic fibroblast growth factor (bFGF), and modulators of the WNT signaling pathway (particularly WNT3A) [8, 9]. These factors were combined to initiate mesendodermal induction, which is the formation of a PS-like population from hPSCs via EMT transition in a gastrulation-like process [10].

To further direct the emerging PS-like cell population toward cardiac mesoderm, inhibition of WNT pathway signaling [7] and the absence of insulin and insulin-like growth factors were found to be required [11–13]. In some studies, supplementation with vascular endothelial growth factor (VEGF), bFGF, and stem cell factor (SCF) at relatively low concentrations at the cardiac mesoderm stage increased cardiac yield [7].

Recent studies indicate that CM subtype-specific differentiation (reviewed by Hausburg et al. [14]) into atrial, ventricular, and pacemaker cells is specified relatively early (i.e., at the cardiac mesoderm stage) [15]. However, most in vitro differentiation protocols apply activators and inhibitors of retinoic acid (RA) signaling to modulate atrial/pacemaker-like and ventricular-like CM specification, respectively [15–18]. Combined with BMP4, RA addition directs the cardiac mesoderm toward pacemaker-like cell [15]. Furthermore, FGF and BMP signaling are required to induce expression of the cardiac transcription factor NKX2.5, which is expressed in ventricular- and atrial-like cells [15, 16, 19].

Recently, insulin growth factor (IGF)-1 was shown to enhance the expansion of CVPs [19], but not differentiated hPSC-CMs at later stages [20]. Similarly, endothelin-1 (EDN1) supported expansion of CVPs in combination with WNT3A [21]. However, "capturing" and expanding cells at the cardiac mesoderm stage or the subsequent CVP stage remains challenging. It was convincingly shown only via the transgene-dependent overexpression of the oncogene c-Myc in CVPs [19]. This process, however, is not straightforward for the envisioned use of CVP-derived CMs for therapies and screening assays. Nevertheless, investigations into the characterization of cells at early cardiac progenitor stages remain highly relevant for both basic research and the applied use of such early, probably proliferative cells for alternative CM production purposes and therapeutic approaches.

2.2   Advantages of Using Small Molecules to Direct Cardiac Cell Fate

In the context of this paper, we define small molecules as pharmaceutically active compounds of low molecular weight, mostly below <1,000 g/mol [22]. In general, these molecules can cross cell membranes via diffusion, a prerequisite for high bioavailability and function; but data on biodistribution, cell penetration, and half-life in vitro are often not available for individual molecules. In the cell, however, small molecules reversibly interact with specific biomolecules, thereby modulating specific signaling pathways and eventually controlling cell phenotype. Varying the concentration of a compound can ideally achieve precise tuning of complex signaling networks. The chemical origin of small molecules offers a toolbox of virtually unlimited diversity for producing compounds to target all kinds of biomolecules and thus direct very distinct biological effects. However, many small molecules interact not only with one cellular target but also with different proteins/pathways and may thus exert unwanted off-target effects, which are often not fully investigated [23, 24].

Because of their relative simple structure, chemical synthesis and bulk production of small molecules is typically straightforward and cost-effective compared with biological substances. Moreover, the production of clinical-grade biologicals such as recombinant growth factors often requires complex bioprocessing and can suffer from batch-to-batch variation, improper biological activity (e.g., from wrong glycosylation and impaired formation of tertiary structures), and relative low half-life in culture media. In consequence, rigorous quality testing of a product's purity and biological activity is required, increasing the overall production costs. Furthermore, the activity and stability of growth factors can vary between culture conditions and media supplements. For example, WNT3A loses its activity in serum-free media unless lipids are added for stabilization [25]. Because cost-effectiveness is an important factor, particularly regarding process upscaling, it is worth highlighting that growth factors and cytokines typically account for the major part of total expenses in bioprocessing, even in small-scale experiments [26].

Consequently, small molecules are promising candidates as effective, stable, and affordable alternatives to biological compounds in regenerative medicine and particularly in production of (stem) cell progeny. During the past decade, a wide range of small molecules have been demonstrated to enhance somatic cell reprogramming into iPSCs, as well as maintenance, expansion, and differentiation of hPSCs (reviewed by Li et al. [27] and Baranek et al. [28]). Regarding cardiac differentiation, several growth factors have been successfully replaced by small molecules. Table 1 provides an overview of key compounds, particularly in the context of cardiac fate modulation.

The initial step of PS induction in hPSCs (marked by the upregulation of T, MIXL, and EOMES) was efficiently induced using the glycogen synthase kinase 3 (GSK3) inhibitor CHIR99021, resulting in WNT pathway activation [1, 30, 33]. Subsequent supplementation with WNT pathway inhibitors such as IWR1, IWP2, WNT-C59, XAV939, and Ky02111 efficiently replaced the secreted WNT antagonist Dickkopf (DKK). As a downstream effect, the above listed factors suppressed the upregulation of transcription factors such as CDX1/2 and MSX1, and thereby modulated cardiac specification [4]. These molecules are now applied in a wide range of cardiac differentiation protocols [36, 37, 42, 72].

Using a monolayer-based approach, Lian et al. reported that the combination of CHIR99021 for 24 h to induce mesendodermal patterning and subsequent specification by IWP2 supplementation (2 days later for 48 h) efficiently generated CMs at relative high purity (>80%) without the addition of any growth factors [30, 73]. Because biphasic control of the WNT signaling pathway is known to play a central role during cardiac development in the embryo and in vitro differentiation of hPSCs, this pathway and its targeting small molecules are outlined in more detail in Sect. 2.2.1.

Table 1 Small molecules applied during cardiac differentiation

Small molecule	Primary target	IC50/EC50	Molecular weight (g/mol)	Molecular effect	Reference	Effect on hPSC differentiation	Reference
BIO	GSK3	5 nM	356.17	WNT pathway activation	Meijer et al. [29]	Induces mesodermal differentiation	Lian et al. [30]; Titmarsh et al. [31]
CHIR99021	GSK3α, GSK3β	10 nM, 6.7 nM	465.34	WNT pathway activation	Ring et al. [32]	Induces mesodermal differentiation; Induces proliferation of CVP's and CMS	Gonzalez et al. [33]; Lian et al. [30]; Titmarsh et al. [20]; Cao et al. [34]
IWR-1	AXIN	180 nM	409.44	WNT pathway inhibitor via axin stabilization	Chen et al. [35]	Induces cardiac specification	Ren et al. [36]; Gonzalez et al. [33]; Hudson et al. [37]
DS-I-7 (IWR-1 analog)	AXIN	4 nM	418.02	WNT pathway inhibitor via axin stabilization	Lanier et al. [38]	Induces cardiac specification	Lanier et al. [38]; Breckwoldt et al. [39]
XAV939	TNKS 1 TNKS 2	11 nM, 4 nM	312.31	WNT pathway inhibitor via axin stabilization	Huang et al. [40]	Induction of cardiac differentiation in ESC	Wang et al. [41]; Minami et al. [42]
IWP-2	PORC	27 nM	466.60	Blocks WNT secretion	Chen et al. [35]	Induces cardiac specification	Lian et al. [30]
WNT-C59	PORC	74 pM	379.45	Blocks WNT secretion	Proffitt et al. [43]	Induces cardiac specification	Burridge et al. [44]
KY02111	Unknown	Not reported	376.855	WNT pathway inhibitor	Minami et al. [42]	Induces cardiac specification	Minami et al. [42]
SB431542	ALK4 ALK5 ALK7	94	384.39	TGF-β inhibitor	Inman et al. [45]	Enhanced cardiac specification	Kattman et al. [46]; Gonzalez et al. [33]
A83-01	ALK5 ALK4 ALK7	12 nM, 45 nM, 7.5 nM	421.52	TGF-β inhibitor; prevents SMAD2/3 phosphorylation	Tojo et al. [47]	Maintenance of CVPs	Chen and Wu [48]; Cao et al. [34]
Dorsomorphin	AMPK ALK2,3,6	109 nM	472.41	BMP inhibitor; prevents SMAD1/5/8 phosphorylation	Yu et al. [49]	Maintenance of CVPs	Cao et al. [34]

Compound	Target	Concentration	Mechanism	Reference	Effect	Reference
DMH-1	ALK2	108 nM	BMP inhibitor; prevents SMAD1/5/8 phosphorylation	Hao et al. [50]	Increased cardiomyogenesis	Aguilar et al. [51]
LDN-193189/DM3189	ALK2 ALK3	5 nM 30 nM	BMP inhibitor; prevents SMAD1/5/8 phosphorylation	Cuny et al. [52]	Blocks mesoderm formation	Loh et al. [53]
Y-27632	ROCK1 ROCK2	220 nM 330 nM	ROCK inhibitor	Uehata et al. [54]	Increases cell viability during differentiation	Fonoudi et al. [55]
Ly294002	PI3K	0.3–6.6 µM	PI3K inhibitor	Vlahos et al. [56]	Blocks cardiac induction, supports definitive endoderm formation	Naito et al. [57]; McLean et al. [58]
PD0325901	MEK	0.33 nM	MEK1/2 inhibitor	Barrett et al. [59]	Inhibits primitive streak induction	Titmarsh et al. [31]
PD173074	FGFR1 FGFR3	5 nM 21.5 nM	FGFR1 and −3 inhibitor	Bansal et al. [60]	Enrichment of SANLPC	Protze et al. [15]
SB203580	p38-α p38-β	300–500 nM	p38 MAPK inhibitor	Cuenda et al. [61]	Enhances cardiac differentiation yield	Graichen et al. [62]
TA-02	p38α CK1ε CK1δ	20 nM 32 nM 32 nM	p38 MAPK inhibitor CK1 inhibitor	Laco et al. [24]	Enhances cardiac differentiation yield	Laco et al. [24]
SAG	SMO	3 nM	SMO receptor agonist	Chen et al. [63]	Supports CVPs maintenance	Birket et al. [19]
Purmorphamine	SMO	2.5 nM	SMO receptor agonist	Wu et al. [64]	Supports induction of NKX2.5	Gonzalez et al. [33]; Fonoudi et al. [55]
IDE1/IDE2	Unknown	125 nM	Induces SMAD2 phosphorylation	Borowiak et al. [65]	Induces endoderm differentiation	Borowiak et al. [65]
AG1478	ErbB1	3 nM	Inhibits ErbB1 (EGFR) signaling	Levitzki and Gazit [66]	Increases fraction of nodal-like CMs	Zhu et al. [67]

(continued)

Table 1 (continued)

Small molecule	Primary target	IC50/EC50	Molecular weight (g/mol)	Molecular effect	Reference	Effect on hPSC differentiation	Reference
1-EBIO	SK1 SK2 SK3 SK4	631 µM 453–866 µM 789–1,040 µM 28.4–100 µM	162.19	Activates SK channels	Devor et al. [68]	Induces shift in subtype composition by depletion of ventricular-like CMs	Jara-Avaca et al. [69]
WY-14643	PPARα	630 nM	323.8	PPARα agonist	Santilli et al. [70]	Induces metabolic maturation of CMs	Poon et al. [71]

AMPK cyclic AMP-dependent protein kinase, *FGFR* fibroblast growth factor receptor, *MEK* mitogen-activated protein kinase kinase, *PI3K* phosphatidylinositol 3-kinase, *PPAR α* peroxisome proliferator-activated receptor alpha, *SK1* small conductance Ca²⁺-activated potassium channel protein 1, *SMO* Smoothened

In addition to modulation of the WNT pathway, inhibition of BMP and TGF-β signaling during cardiac specification was reported using DMH-1 [51] and SB431542 [33, 46], respectively. Furthermore, tuning mitogen-activated protein kinase (MAPK) activity by SB203580 was shown to improve early induction of differentiation [74] and increase CM yield [62]. Interestingly, recent studies reported inhibition of the WNT pathway by SB203580 and its analog TA-02 via inhibition of CK1 [24], suggesting multiple roles of SB203580 during the cardiac differentiation process. Inhibition of ErbB signaling using the ErbB1 (EGFR) antagonist AG1478 increased the fraction of the nodal-like CM subtype [67].

Regarding maintenance of CVPs, Cao et al. reported the expansion of SSEA1$^+$ CVPs solely based on the combination of small molecules CHIR99021, dorsomorphin (a BMP pathway antagonist), and A83-01 [34]. Recently, the Smoothened agonist SAG in combination with IGF-1 was also shown to support CVP maintenance [19]. Interestingly, purmorphamine, another Smoothened agonist, was used to drive cardiac specification toward NKX2.5-positive cells [33, 55].

Most small molecules in cardiac differentiation protocols interfere with kinase activity in major signaling pathways, but other classes of molecules have also been described. For example, 1-EBIO, a positive modulator of Ca^{2+}-activated K^+ channels, was reported to induce cardiac differentiation in murine PSCs, thereby directing specification into CMs with pacemaker-like properties [75]. However, this inductive effect was not confirmed using human PSCs [69]. Instead, treatment with 1-EBIO resulted in selective survival at the cardiac progenitor stage by depleting proliferative, noncardiac cells and reducing the proportion of CMs with ventricular-like properties. Ultimately, this resulted in the enrichment of CMs with shortened actin potentials, reminiscent of nodal- and atrial-like phenotypes [69].

2.2.1 Small Molecules Regulating WNT Pathway Activity

The WNT signaling pathway is a complex and dynamic signaling network involved in the regulation of fundamental cellular processes in the embryonic state, including self-renewal, proliferation, cell fate determination, and differentiation [76]. β-Catenin is a highly conserved, central transcriptional effector of this signaling cascade (commonly referred to as the "canonical" WNT signaling pathway). In the absence of WNT ligands, β-catenin is phosphorylated and recognized by the E3 ubiquitin-ligase complex, resulting in ubiquitination and degradation via the 26S proteasome pathway (Fig. 3a). The phosphorylation of β-catenin is mediated by GSK3 and casein kinase 1α (CK1), with CK1 phosphorylating at serine-45 (S45) to prime β-catenin for subsequent GSK3 phosphorylation at serine-33/serine-37/threonine-41 (S33/S37/T41). This process is mediated by the β-catenin destruction complex consisting of β-catenin, CK1, GSK3, and the scaffold proteins Axin and adenomatous polyposis coli (APC).

Upon activation of canonical WNT signaling (Fig. 3b), secreted WNT ligands bind to the transmembrane receptor Frizzled (Fz) and low-density lipoprotein receptor-related protein coreceptor 5/6 (LRP5/6), thus initiating inhibition of the

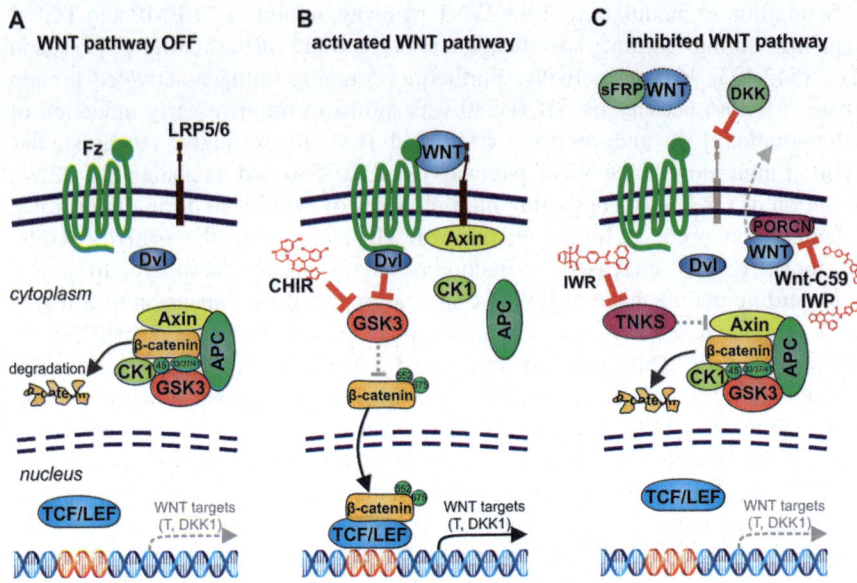

Fig. 3 WNT signaling pathway. (**a**) In the absence of WNT pathway agonists, β-catenin is sequestrated by the β-catenin destruction complex and degraded by the proteasome in a phosphorylation-dependent process. (**b**) Binding of WNT to Fz and LRP5/6 results in inhibition of the β-catenin destruction complex and stabilization of β-catenin. β-Catenin translocates into the nucleus, interacts with TCF/LEF, and thereby regulates transcription of WNT target genes. Alternatively, CHIR results in WNT activation by inhibition of GSK3. (**c**) Mechanisms of different classes of WNT inhibiting molecules such as sFRP, DKK, and the small molecules IWR and IWP. Key phosphorylation sites of β-catenin are indicated by *green circles*

β-catenin destruction complex. In this process, Dishevelled (DVL) recruits rate-limiting Axin to the cell membrane and destabilizes the complex. Unphosphorylated β-catenin (at S33/S37/T41 and S45) escapes degradation, accumulates in the cytoplasm, and translocates into the nucleus [77]. In the nucleus, β-catenin interacts with TCF/LEF transcription factors, guiding β-catenin to specific WNT response elements (WRE) on target genes, where β-catenin acts as central transcriptional activator [78]. The list of known downstream targets regulated by β-catenin (dependent on the cell type and state) comprises more than 100 genes including NANOG, OCT3/4, T, ISL1, SOX2, SOX9, SOX17, CDX1, TNFRSF19, JUN, BMP4, WNT3A, FGF4, FGF9, FGF18, CDH1, AXIN2, DKK1, and SFRP2 [79].

The canonical WNT signaling pathway is tightly regulated at various levels, including several positive and negative regulatory mechanisms. Enhanced pathway activation results from phosphorylation of β-catenin at serine-675 via protein kinase A (PKA) as well as phosphorylation at serine-552 via PKA and AKT [80–82]. Inhibitory mechanisms include the inhibitor of β-catenin and TCF-4 (ICAT), which inhibits formation of a complex of β-catenin with TCF4 and functions as a buffer at increased levels of β-catenin [83]. Secreted frizzled-related protein (sFRP) and

WNT inhibitory factor (WIF) directly bind to WNTs and prevent their interaction with Fz. Furthermore, DKK interacts with the LRP5/6 receptor and induces internalization of the coreceptor, resulting in decreased WNT signaling activity [76].

As outlined in Table 1, several small molecules can modulate WNT pathway activity. A highly potent agonist of the WNT pathway is the small molecule aminopyrimidine CHIR99021 (CHIR), which has been extensively used in hPSC research in recent years. Inhibition of GSK3 causes decreased phosphorylation of β-catenin at S33/S37/T41, resulting in stabilization of β-catenin and consequent activation of the WNT pathway. CHIR is widely applied in stem cell research for maintenance of mouse PSCs [84] and the induction of differentiation of hPSCs [85] toward numerous mesendodermal lineages such as hepatocytes [86], CMs [1, 33], endothelial cells [87], renal like-cells [88, 89], smooth muscle cells [87], skeletal muscle cells [90], and chondrogenic cells [3].

A common target of small molecules with inhibitory effects on the WNT pathway (Fig. 3c) is the group of Tankyrase (TNKS) inhibitors such as IWR1 and XAV939. TNKS is thought to regulate AXIN degradation. AXIN is stabilized upon inhibition of its activity, resulting in sequestration of β-catenin in the destruction complex and subsequent WNT pathway inhibition [91]. Another group of WNT inhibitors, acting on WNT-producing cells, includes inhibitors of porcupine (PORC) such as IWP and WNT-C59. PORC is an acyltransferase in the endoplasmic reticulum that palmitoylates WNTs, a prerequisite for WNT secretion. By blocking palmitoylation of WNTs, secretion of WNTs is blocked and signaling is suppressed.

As noted above, it is remarkable how many functional cell types of mesendodermal origin were successfully derived from hPSCs using a small subset of chemical WNT modulators, in particular CHIR. Even more surprising is the use of highly equivalent and even overlapping concentrations of CHIR compounds in individual protocols aimed at directing hPSCs into entirely different functional lineages (e.g., CMs versus hepatocytes). Section 3 discusses recent work by our group and others that addresses this aspect, which is of central interest for the development of large-scale robust and reproducible differentiation processes for production of specific cell types.

3   Trends, Advanced Technologies, and Strategies

Given the significant irreversible loss of human CMs induced by ischemic cardiomyopathies, it has been estimated that billions of in vitro-derived CMs per patient are required to compensate for this loss by cell therapy aiming to save the diseased heart from failure [92, 93].

To avoid immunological barriers, initial clinical trials may aim at the production of individual, patient-specific cell batches derived from patients' own iPSCs. However, in this scenario, extensive quality control procedures and safety measures have to be performed for each individual cell batch. This is extremely time-, labor-

and cost-intensive, thus challenging the envisioned broad applicability of hPSC-based therapies [94].

Instead, the field aims at developing allogeneic approaches using a conventional immunosuppression regimen, alone or in combination with "universal" human leucocyte antigens (HLA)-matched or depleted hPSC lines [95]. The allogeneic strategy allows (and requires) generation of large batches of cells such as CMs, which can be extensively characterized, stored, and serve as an "of-the-shelf product on demand" in the clinic. Notably, the first transplantation of PSC progenies into a human heart was recently performed by an allogeneic approach using an hESC line as cell source [96].

For large screening of extensive compound libraries (often consisting of >10,000 or even >100,000 samples), pharma companies also require well-defined batches of functional cells such as hPSC-CMs. This is also important for reproducing and validating relevant data at later states of the drug discovery process. Thus, routine therapies as well as industry-compliant screening approaches require efficient and highly reproducible processes for the mass production of functional CM batches. We have recently reviewed general progress in the fields of process development for hPSC production and cardiac differentiation in scalable bioreactor systems [97, 98].

In the following section, we focus on the required media formulations, a key component in process development.

3.1 Defined Culture Media for Differentiation and Mass Production

Murine embryonic stem cells (mESC) were described in the early 1980s [99] and were thus available long before human ESCs and iPSCs; thus, the mESC model was extensively used for developing culture processes for PSC expansion, differentiation, and scale-up.

For cardiomyogenic differentiation of mESCs, dynamic conditions (i.e., rotation of cell aggregates in suspension in conventional culture dishes) support culture homogeneity, process reproducibility, and cardiac differentiation when ultimately combined with genetic lineage selection [100]. This strategy paved the way for mESC-CM production in stirred flasks [101], subsequent transition into fully monitored and controlled stirred tank reactors at the 2 L scale [102], and process optimization [103]. However, in addition to the heterogeneity of embryoid bodies (EBs; aggregates of PSCs primed for differentiation) that are randomly formed in suspension [104, 105], the process outcome strongly depends on the applied batch of fetal calf serum (FCS) as an essential component of the culture medium. One of the first reports on cardiomyogenic differentiation of hESCs demonstrated the induction of contracting CMs in EBs by applying differentiation medium supplemented with 20% FCS [106]. The CM induction efficiency achieved in such medium, however, was extremely low (<1% CMs).

In the search for chemically defined and more cost-efficient differentiation media, work by our group and others revealed that the omission of mitogenic stimuli such as FCS and insulin at the first stage(s) of differentiation (i.e., during the first 3 days of hPSC specification into a PS-like stage and precardiac mesoderm) substantially promoted CM induction [11, 13]. These results, together with the finding that the p38 MAPK inhibitor SB203580 induces cardiogenesis, led to development of the first chemically defined cardiac differentiation medium for hESCs, termed "basic serum-free" (bSF) medium [13, 62]. Optimization of this differentiation strategy enabled the induction of up to 10–20% CMs (depending on the applied hESC line) when differentiation was performed on surface-attached cells in two dimensions (2D). Moreover, combining the method with genetic lineage enrichment resulted in essentially 100% human CM purity [107, 108].

The chemically defined differentiation medium was subsequently used for differentiation of mass-produced hESC aggregates grown in stirred suspension culture in flasks and bioreactors [109, 110]. However, in contrast to the successful differentiation outcome achieved on surface-attached cells in 2D, supplementation of SB203580 in bSF medium to hESC aggregates in suspension resulted in quantitative loss of viable cells instead of cardiogenic differentiation (Kempf et al., unpublished). These observations suggest that the culture platform applied at the pluripotent state (the "preculture" ahead of differentiation) substantially impacts subsequent differentiation results. Interestingly, when hESCs or hiPSCs were expanded in medium preconditioned with mouse embryonic fibroblasts (MEFs), the resulting EBs complied with differentiation in bSF medium in suspension culture, at least in stirred culture dishes at typical laboratory scale [111]. Genetically enriched "cardiac bodies" (contracting aggregates consisting of essentially pure CMs derived by this method) were successfully used to generate force-generating bioartificial cardiac tissue (BCT) [112]. However, the dependence on poorly defined MEF-conditioned medium for hPSC expansion is problematic with respect to process upscaling at chemically defined conditions.

As outlined above, heart development during differentiation depends on biphasic WNT pathway modulation, comprised of upregulation at an early PS-formation stage and downregulation at a later CM specification step [113]. Following this idea, protocols were developed using chemical WNT modulators to mimic this biphasic pattern. CHIR was used for WNT induction and IWP or IWR for WNT pathway suppression [30, 33], as extensively discussed above. Although the strategy was successful on monolayer cultures of hPSCs expanded in the mTeSR medium on Matrigel-coated dishes, we showed that this protocol can be successfully adapted to matrix-free hPSC aggregates cultured in suspension. After process optimization in multiwell dishes and agitated Erlenmeyer flasks, successful transition into impeller-stirred tank bioreactors was possible, generating ~40 million human CMs in a 100 -mL scale process at ~80–90% CM purity, independent of additional enrichment steps [114, 115]. Notably, specific conditions for the generation of aggregates at the pluripotent state were required for successful cardiac differentiation, further highlighting the important role of preculture.

Focusing on this topic, we systematically investigated how the "bulk cell density" (BCD), defined as the number of cells in a given culture volume, affects the differentiation result [1]. In addition to the known effect of the concentration of the WNT modulator CHIR, an unexpected and poorly appreciated effect of the BCD in combination with the CHIR concentration was observed (Fig. 1). We demonstrated that, within the first 24 h of differentiation induction, hPSCs secrete a complex mixture of factors into the medium, many of which are known modulators of developmental processes, including BMP and WNT pathway agonist/antagonists. Secreted candidates were identified via gene expression arrays of hPSCs and mass spectrometry analysis of conditioned media. Respective candidates were then tested either by the supplementation of recombinant protein to the medium or by gene knockdown experiments that modulated the accumulation of relevant factors during early stages of differentiation. We identified two specific molecules, TGF-β family members LEFTY1 and CER1 (regulators of the BMP/Nodal pathway), as specific modulators of PS-like priming in hPSCs, which consequently dictates the formation of mesendodermal and, thus, cardiac progenies at later stages of differentiation. This work reveals how specific factors such as LEFTY1, which is readily expressed in hPSCs at the pluripotent state, modulate subsequent differentiation. Furthermore, these finding highlight how important it is to closely monitor and control the BCD with respect to the robustness and reproducibility of hPSC differentiation. This topic is particularly relevant to process development and upscaling in suspension culture, where the cell density needs to be defined by the investigator instead of being "automatically restricted" by the available cell surface in a culture dish in 2D.

These studies mainly relied on mTeSR medium for hPSC cultivation. However, more recent studies revealed that this rather complex medium can be replaced by the simpler formulation E8 for hPSC expansion in stirred suspension. Importantly, in principle, E8 can be generated from chemically defined components. Using both mTeSR and E8, we recently showed that the typical fed-batch cultivation (defined by "all-in-one" daily medium replacement in 2D and 3D) can be replaced by perfusion feeding (achieved by a constant medium flow-through) in specific stirred bioreactor systems. Cell retention systems were used to avoid perfusion-induced cell loss [116]. Constant medium perfusion avoids the typical zig-zag pattern of process parameters such as glucose, lactate, and pH, resulting in more homogeneous culture conditions. Despite the same medium consumption (compared with the parallel repeated batch feeding strategy), homogenous culture conditions readily induce ~50% increased cell yields independent of the applied cell lines (hESC or hiPSC) or expansion medium [116].

It is worth noting that, for cardiac differentiation, the complex media supplement B27 is extensively used [6], often in combination with RPMI1640 basal medium. In the search for a replacement for this multicomponent and expensive formulation, a number of alternatives have been described in the literature. The Greber group carried out systematic combinatorial optimization of medium components, resulting in a simple serum and serum albumin-free basal medium in which cardiomyogenesis is mediated by a minimal set of signaling pathway manipulations at moderate factor concentrations [117]. Moreover, Burridge et al. published a chemically defined cardiac differentiation medium termed CDM3 composed of only three ingredients (RPMI1640, human serum albumin, and ascorbic acid). The "minimalistic approaches" by Greber and Burridge were pushed to the edge by using plain RPMI1640 for differentiation [118]. Although this may work in principle in the context of surface-attached 2D differentiation, our preliminary results strongly suggest the incompatibility of this lean medium for CM differentiation in stirred suspension culture (Halloin et al., unpublished). On the other hand, our preliminary results suggest the utility of CDM3 for cardiac differentiation in stirred bioreactors (Halloin et al. unpublished). Combined with hPSC aggregate expansion in E8, this opens the perspective for chemically defined and commercially viable CM production processes at the multiliter scale.

In summary, 16 years after the first publication describing the formation of contracting CMs from human ESCs [106], followed by the derivation of cardiac and other lineages from the first clinical grade hESC lines in 2006 [119] and cord blood-derived hiPSCs in 2009 [120], the mass production of human CMs in defined media at the 100 mL [114, 115] and 1 L process scales [121] has been achieved.

Given the multiple variables that impact on lineage induction and specification, process adaptation to individual hPSC lines and interprocess variability are still major issues in the field regarding the percentage of CM induction and the overall cell yield. With the possibilities for feedback loop-based process control in stirred fully instrumented bioreactors, it is expected that substantial process improvement and cell line-independent process reproducibility will be achieved.

3.2 Inducing Maturation by Tissue Engineering and Metabolic Strategies

In addition to the challenge of producing hPSC-derived CMs, their lack of maturity is another major problem in the field and the subject of controversy with respect to their envisioned use for heart repair.

On the one hand, the endogenous automaticity of hPSC-CMs is a key issue for therapies. Despite their often ventricular-like molecular and electrophysiological features, hPSC-CMs typically display an autonomous nonpacemaker-dependent membrane depolarization and subsequent contractile ability, which is not typical for quiescent CMs in the working myocardium of the adult heart. As demonstrated in a nonhuman primate model, the transplantation of hESC-CMs results in the induction of arrhythmias, a major safety concern in translation medicine [122]. The underlying mechanisms of arrhythmia induction have not been fully resolved, but it is suspected that the automaticity and/or the lack of functional coupling of the transplanted hPSC-derived donor CMs with the host tissue play a role. However, the presence of hPSC-derived nodal/pacemaker-like CMs in the transplanted cell population may also have an effect.

On the other hand, the lack of CM maturity might support hPSC-CMs engraftment in the heart because they may better tolerate hypoxia, have a higher structural plasticity, and maintain some proliferative potential after engraftment, in contrast to fully matured cells. Moreover, transplantation at an even earlier cardiac progenitor state (CVPs) may support the ability to form multiple cardiac cell types in situ, including endothelial cells and smooth muscle cells in addition to CMs, which could be an advantage for effective therapies [96, 123].

In contrast, for in vitro disease modeling, tissue engineering, and drug screening assays, induction of the highest possible degree of CM maturation in a dish is desirable. Apparently, maturation can (and must) be assessed at the following levels:

1. Expression patterns of microRNAs/long-noncoding RNAs, genes, and proteins
2. Electrophysiological features such as ion channel patterns, presence and strength of ion currents, and the resulting overall action potential (AP) or field potential (FP) patterns
3. Response to pharmacological inhibitors/stimulators of cardiac properties
4. Calcium handling properties
5. Metabolic maturation
6. Structural maturation such as T-tubule formation, striation of sarcomeric structures, and (ultimately) contractile force formation

All of these features are mutually dependent but it is well established that typical hPSC-CMs at ~2 weeks after hPSC differentiation have an early embryonic phenotype rather than fetal, postnatal, or even adult phenotype [124, 125]. Moreover, most strategies applied to date have achieved CM maturation regarding some of the aspects outlined above, but limited progress toward fully mimicking the adult CM phenotype.

In general, two basic principle of in vitro maturation have been applied: (1) tissue engineering, often including mechanical and/or electrical stimulation and (2) supplementation of chemical, pharmacological, and metabolic substances aiming at stimulating maturation by paracrine cues. Apparently, none of these strategies is mutually exclusive.

Recent work by our group [112] and others [39, 126, 127] have demonstrated that tissue engineering, which typically combines CMs with other cell types (such as matrix forming fibroblast) and/or decellularized matrices [128] and exposes the resulting constructs to mechanical stimulation, induces many aspects of cardiac tissue maturation such as more pronounced and ordered sarcomere structures, stronger contractile forces, and improved electrophysiological coupling [129].

Irrespective of advanced tissue engineering, our group [130] and others [131] have recently shown that the simple seeding and long-term incubation of CMs (i.e., up to 100 days post-differentiation) on stiff surfaces such as glass [130] or patterned surfaces displaying specific grooves and ridges [132] can induce relevant aspects of CM maturation at the single cell level. For example, long-term cultivation induced improved mitochondrial maturation in terms of increased mitochondrial relative abundance, enhanced membrane potential, and increased activity of several mitochondrial respiratory complexes [133].

Focusing on sarcomere maturation, we observed an entire switch from embryonic/fetal isoform expression of sarcomere proteins such as alpha myosin heavy chain (αMHC, MYH6) and myosin light chain 2 atrial (MLC2a, MYL7) toward isoforms typical of the adult heart [i.e., beta MHC (βMHC, MYH7) and MLC 2 ventricular (MLC2v, MYL2), respectively] [130]. Although long-term cultivation had a limited impact on the twitch kinetics and electrophysiological properties of cells, the quantitative switch toward mature sarcomere protein isoforms was very valuable for studying the role of adult protein isoforms in hPSC-CMs. A typical example is familial cardiac hypertrophy (FCH), the most common genetically induced form of heart failure, caused by inherited mutations in the βMHC gene (MYH7) [134]. In vitro disease models of FCH rely on the quantitative induction of βMHC expression (in exchange for αMHC) in patient-specific hiPSC-CMs to enable the induction of a potential disease phenotype in a dish. In a first in vitro model of FCH, some disease aspects (in particular improper calcium handling) were suggested without achieving a proper switch from α-to-βMHC isoform expression. Which phenotypic properties of FCH patient-derived hiPSC-CMs will be displayed by cells that exclusively express the βMHC isoform still needs elucidating.

In the next part of this section, we highlight some recent publications that report supplementing the medium with specific compounds to induce "adolescence" in hPSC-CMs.

Although most cell types in the body rely on glucose as the major carbon source for their energy metabolism, CMs are more flexible. There are distinct changes between energy substrate utilization by CMs before and after birth. Upon development and maturation, CMs switch from glucose consumption in the embryonic state to the use of fatty acids as the main energy source in the adult heart. However, in the transition state, the fetal heart uses lactate as the major energy source for ATP production. Taking advantage of these metabolic properties, Tohyama and coworkers replaced glucose by lactate in a medium for hPSC-CMs cultivation after differentiation. Because the non-CMs present upon differentiation did not tolerate glucose deprivation, the medium enabled enrichment to almost pure CM populations [135]. In our hands, however, we observed a substantial loss of overall cell numbers upon the addition of lactate-based medium after cardiac differentiation of hPSCs. Notably, this cell loss included the substantial depletion of CMs (Kempf et al., unpublished), which tallies with the observation by other investigators [136]. These results suggest, again, that early derived CMs have an embryonic rather that fetal phenotype, including their metabolism, and therefore only a subpopulation can tolerate the "lactate-only diet."

More recently, Drawnel et al. used hiPSC-CMs for the in vitro modeling of diabetic cardiomyopathy (DCM) as a complication of type 2 diabetes [137]. Aiming at promoting adult patterns of metabolic activity, a maturation medium was introduced containing insulin and fatty acids, but no glucose, for 3 days after plating of early CMs. Although a full adult phenotype was not attained, metabolic dependence on fatty acid β-oxidation supported cellular activities prominent in more mature cells with respect to structural, molecular, and electrophysiological features. Notably, no increase in the proportion of dead cells was observed in the glucose-free medium, suggesting that the applied maturation medium was not selecting against an immature cell population but was actively promoting CM maturation.

In another study, CMs were exposed to a one-week treatment with the growth hormone tri-iodo-L-thyronine (T3), which is known to be essential for optimal heart growth [138]. Analysis of an array of morphological, molecular, and functional parameters demonstrated that T3 drives some degree of hiPSC-CM maturation compared with the untreated control and may enhance their utility for assays. It remains to be tested whether the combination of T3 with parallel or serial supplementation, for example, with glucose-free treatments can induce an even more pronounced maturation phenotype in vitro.

4   Conclusion and Outlook

As outlined above, substantial progress has been recently achieved regarding the production and application of hPSC-CMs. The development of defined culture media and supplementation with both naturally occurring factors and chemical pathway modulators were successfully utilized to fuel this progress. Although recombinant factors directing differentiation and maturation are extremely valuable for mimicking conditions known from developmental biology, the use of chemically synthesized compounds seems favorable for envisioned routine mass production of cells.

With respect to the in vitro maturation of CMs, major progress has been achieved, mainly using a complex tissue engineering approach and more simple mechanical tricks such as cell seeding on matrices with specific rigidity or patterning, followed by prolonged cultivation. However, metabolic engineering by feeding CMs with alternative carbon sources such as lactate and fatty acids seems extremely promising for mass production of CMs in suspension culture.

For generation of cardiac tissue-like structures compliant with high-throughput screening, the generation of multicellular spherical organoids composed of organ-specific cell types is an upcoming technology currently swamping the field as recently reviewed [139, 140], with the first examples being reported in the cardiac field [141, 142].

However, for many hiPSC-CM-based in vitro models, in particular for inherited electrophysiological phenotypes [143], the relative simple use of seeded CMs followed by treatment with pharmacological ion channel modulators and patch clamp or microelectrode analysis seems to be highly informative [144]. Moreover, even monitoring patient-specific sensitivity of CMs to chemicals such as the chemotactic anticancer drug doxorubicin is informative and simply involves derivation of hiPSC-CMs specific to the patient without requiring advanced cardiac tissue engineering [145].

Another challenge is the efficient transplantation and functional integration of hPSC-CMs for therapeutic heart repair in situ. Here, the field is in full swing, testing and comparing different methods of cell transplantation starting from single cells up to complex engineered patches. Formation of injectable, multicellular microtissue presents an interesting alternative that could support both immediate cell retention after cell administration and long-term engraftment [146–148]. These options are currently being tested in physiologically relevant models by us, with national (iCARE) and trans-European (TECHNOBEAT) collaboration.

References

1. Kempf H, Olmer R, Haase A, Franke A, Bolesani E, Schwanke K, Robles-Diaz D, Coffee M, Gohring G, Drager G, Potz O, Joos T, Martinez-Hackert E, Haverich A, Buettner FF, Martin U, Zweigerdt R (2016) Bulk cell density and Wnt/TGFbeta signalling regulate mesendodermal patterning of human pluripotent stem cells. Nat Commun 7:13602. https://doi.org/10.1038/ncomms13602
2. Faial T, Bernardo AS, Mendjan S, Diamanti E, Ortmann D, Gentsch GE, Mascetti VL, Trotter MW, Smith JC, Pedersen RA (2015) Brachyury and SMAD signalling collaboratively orchestrate distinct mesoderm and endoderm gene regulatory networks in differentiating human embryonic stem cells. Development 142(12):2121–2135. https://doi.org/10.1242/dev.117838
3. Mendjan S, Mascetti VL, Ortmann D, Ortiz M, Karjosukarso DW, Ng Y, Moreau T, Pedersen RA (2014) NANOG and CDX2 pattern distinct subtypes of human mesoderm during exit from pluripotency. Cell Stem Cell 15(3):310–325. https://doi.org/10.1016/j.stem.2014.06.006
4. Rao J, Pfeiffer MJ, Frank S, Adachi K, Piccini I, Quaranta R, Arauzo-Bravo M, Schwarz J, Schade D, Leidel S, Scholer HR, Seebohm G, Greber B (2016) Stepwise clearance of repressive roadblocks drives cardiac induction in human ESCs. Cell Stem Cell 18(3):341–353. https://doi.org/10.1016/j.stem.2015.11.019
5. Wu Q, Zhang L, Su P, Lei X, Liu X, Wang H, Lu L, Bai Y, Xiong T, Li D, Zhu Z, Duan E, Jiang E, Feng S, Han M, Xu Y, Wang F, Zhou J (2015) MSX2 mediates entry of human pluripotent stem cells into mesendoderm by simultaneously suppressing SOX2 and activating NODAL signaling. Cell Res 25(12):1314–1332. https://doi.org/10.1038/cr.2015.118
6. Laflamme MA, Chen KY, Naumova AV, Muskheli V, Fugate JA, Dupras SK, Reinecke H, Xu C, Hassanipour M, Police S, O'Sullivan C, Collins L, Chen Y, Minami E, Gill EA, Ueno S, Yuan C, Gold J, Murry CE (2007) Cardiomyocytes derived from human embryonic stem cells in pro-survival factors enhance function of infarcted rat hearts. Nat Biotechnol 25(9):1015–1024. https://doi.org/10.1038/nbt1327
7. Yang L, Soonpaa MH, Adler ED, Roepke TK, Kattman SJ, Kennedy M, Henckaerts E, Bonham K, Abbott GW, Linden RM, Field LJ, Keller GM (2008) Human cardiovascular progenitor cells develop from a KDR+ embryonic-stem-cell-derived population. Nature 453(7194):524–528. https://doi.org/10.1038/nature06894
8. Leschik J, Stefanovic S, Brinon B, Puceat M (2008) Cardiac commitment of primate embryonic stem cells. Nat Protoc 3(9):1381–1387. https://doi.org/10.1038/nprot.2008.116
9. Tran TH, Wang X, Browne C, Zhang Y, Schinke M, Izumo S, Burcin M (2009) Wnt3a-induced mesoderm formation and cardiomyogenesis in human embryonic stem cells. Stem Cells 27(8):1869–1878. https://doi.org/10.1002/stem.95

10. D'Amour KA, Agulnick AD, Eliazer S, Kelly OG, Kroon E, Baetge EE (2005) Efficient differentiation of human embryonic stem cells to definitive endoderm. Nat Biotechnol 23 (12):1534–1541. https://doi.org/10.1038/nbt1163

11. Freund C, Ward-van Oostwaard D, Monshouwer-Kloots J, van den Brink S, van Rooijen M, Xu X, Zweigerdt R, Mummery C, Passier R (2008) Insulin redirects differentiation from cardiogenic mesoderm and endoderm to neuroectoderm in differentiating human embryonic stem cells. Stem Cells 26(3):724–733. https://doi.org/10.1634/stemcells.2007-0617

12. Lian X, Zhang J, Zhu K, Kamp TJ, Palecek SP (2013) Insulin inhibits cardiac mesoderm, not mesendoderm, formation during cardiac differentiation of human pluripotent stem cells and modulation of canonical Wnt signaling can rescue this inhibition. Stem Cells 31(3):447–457. https://doi.org/10.1002/stem.1289

13. Xu XQ, Graichen R, Soo SY, Balakrishnan T, Bte Rahmat SN, Sieh S, Tham SC, Freund C, Moore J, Mummery C, Colman A, Zweigerdt R, Davidson BP (2008) Chemically defined medium supporting cardiomyocyte differentiation of human embryonic stem cells. Differentiation 76(9):958–970. https://doi.org/10.1111/j.1432-0436.2008.00284.x

14. Hausburg F et al (2017) Specific cell (re-)programming: approaches and perspectives. doi: https://doi.org/10.1016/j.addr.2017.09.005

15. Protze SI, Liu J, Nussinovitch U, Ohana L, Backx PH, Gepstein L, Keller GM (2017) Sinoatrial node cardiomyocytes derived from human pluripotent cells function as a biological pacemaker. Nat Biotechnol 35(1):56–68. https://doi.org/10.1038/nbt.3745

16. Devalla HD, Schwach V, Ford JW, Milnes JT, El-Haou S, Jackson C, Gkatzis K, Elliott DA, Chuva de Sousa Lopes SM, Mummery CL, Verkerk AO, Passier R (2015) Atrial-like cardiomyocytes from human pluripotent stem cells are a robust preclinical model for assessing atrial-selective pharmacology. EMBO Mol Med 7(4):394–410. 10.15252/emmm. 201404757

17. Pei F, Jiang J, Bai S, Cao H, Tian L, Zhao Y, Yang C, Dong H, Ma Y (2017) Chemical-defined and albumin-free generation of human atrial and ventricular myocytes from human pluripotent stem cells. Stem Cell Res 19:94–103. https://doi.org/10.1016/j.scr.2017.01.006

18. Zhang Q, Jiang J, Han P, Yuan Q, Zhang J, Zhang X, Xu Y, Cao H, Meng Q, Chen L, Tian T, Wang X, Li P, Hescheler J, Ji G, Ma Y (2011) Direct differentiation of atrial and ventricular myocytes from human embryonic stem cells by alternating retinoid signals. Cell Res 21 (4):579–587. https://doi.org/10.1038/cr.2010.163

19. Birket MJ, Ribeiro MC, Verkerk AO, Ward D, Leitoguinho AR, den Hartogh SC, Orlova VV, Devalla HD, Schwach V, Bellin M, Passier R, Mummery CL (2015) Expansion and patterning of cardiovascular progenitors derived from human pluripotent stem cells. Nat Biotechnol 33(9):970–979. https://doi.org/10.1038/nbt.3271

20. Titmarsh DM, Glass NR, Mills RJ, Hidalgo A, Wolvetang EJ, Porrello ER, Hudson JE, Cooper-White JJ (2016) Induction of human iPSC-derived cardiomyocyte proliferation revealed by combinatorial screening in high density microbioreactor arrays. Sci Rep 6:24637. https://doi.org/10.1038/srep24637

21. Soh BS, Ng SY, Wu H, Buac K, Park JH, Lian X, Xu J, Foo KS, Felldin U, He X, Nichane M, Yang H, Bu L, Li RA, Lim B, Chien KR (2016) Endothelin-1 supports clonal derivation and expansion of cardiovascular progenitors derived from human embryonic stem cells. Nat Commun 7:10774. https://doi.org/10.1038/ncomms10774

22. Kalesse M (2014) Introduction to biological and small molecule drug research and development: theory and case studies. Edited by C. Robin Ganellin, Roy Jefferis and Stanley M. Roberts. ChemMedChem 9(4):856–856. https://doi.org/10.1002/cmdc.201300492

23. Breinig M, Klein FA, Huber W, Boutros M (2015) A chemical-genetic interaction map of small molecules using high-throughput imaging in cancer cells. Mol Syst Biol 11(12):846. 10.15252/msb.20156400

24. Laco F, Low JL, Seow J, Woo TL, Zhong Q, Seayad J, Liu Z, Wei H, Reuveny S, Elliott DA, Chai CL, Oh SK (2015) Cardiomyocyte differentiation of pluripotent stem cells with

SB203580 analogues correlates with Wnt pathway CK1 inhibition independent of p38 MAPK signaling. J Mol Cell Cardiol 80:56–70. https://doi.org/10.1016/j.yjmcc.2014.12.003

25. Tuysuz N, van Bloois L, van den Brink S, Begthel H, Verstegen MM, Cruz LJ, Hui L, van der Laan LJ, de Jonge J, Vries R, Braakman E, Mastrobattista E, Cornelissen JJ, Clevers H, Ten Berge D (2017) Lipid-mediated Wnt protein stabilization enables serum-free culture of human organ stem cells. Nat Commun 8:14578. https://doi.org/10.1038/ncomms14578

26. Zaret KS (2009) Using small molecules to great effect in stem cell differentiation. Cell Stem Cell 4(5):373–374. https://doi.org/10.1016/j.stem.2009.04.012

27. Li W, Jiang K, Ding S (2012) Concise review: a chemical approach to control cell fate and function. Stem Cells 30(1):61–68. https://doi.org/10.1002/stem.768

28. Baranek M, Belter A, Naskret-Barciszewska MZ, Stobiecki M, Markiewicz WT, Barciszewski J (2017) Effect of small molecules on cell reprogramming. Mol BioSyst 13 (2):277–313. https://doi.org/10.1039/c6mb00595k

29. Meijer L, Skaltsounis AL, Magiatis P, Polychronopoulos P, Knockaert M, Leost M, Ryan XP, Vonica CA, Brivanlou A, Dajani R, Crovace C, Tarricone C, Musacchio A, Roe SM, Pearl L, Greengard P (2003) GSK-3-selective inhibitors derived from Tyrian purple indirubins. Chem Biol 10(12):1255–1266

30. Lian X, Hsiao C, Wilson G, Zhu K, Hazeltine LB, Azarin SM, Raval KK, Zhang J, Kamp TJ, Palecek SP (2012) Robust cardiomyocyte differentiation from human pluripotent stem cells via temporal modulation of canonical Wnt signaling. Proc Natl Acad Sci U S A 109(27): E1848–E1857. https://doi.org/10.1073/pnas.1200250109

31. Titmarsh DM, Hudson JE, Hidalgo A, Elefanty AG, Stanley EG, Wolvetang EJ, Cooper-White JJ (2012) Microbioreactor arrays for full factorial screening of exogenous and para-crine factors in human embryonic stem cell differentiation. PLoS One 7(12):e52405. https://doi.org/10.1371/journal.pone.0052405

32. Ring DB, Johnson KW, Henriksen EJ, Nuss JM, Goff D, Kinnick TR, Ma ST, Reeder JW, Samuels I, Slabiak T, Wagman AS, Hammond ME, Harrison SD (2003) Selective glycogen synthase kinase 3 inhibitors potentiate insulin activation of glucose transport and utilization in vitro and in vivo. Diabetes 52(3):588–595

33. Gonzalez R, Lee JW, Schultz PG (2011) Stepwise chemically induced cardiomyocyte specification of human embryonic stem cells. Angew Chem Int Ed Engl 50(47):11181–11185

34. Cao N, Liang H, Huang J, Wang J, Chen Y, Chen Z, Yang HT (2013) Highly efficient induction and long-term maintenance of multipotent cardiovascular progenitors from human pluripotent stem cells under defined conditions. Cell Res 23(9):1119–1132. https://doi.org/10.1038/cr.2013.102

35. Chen B, Dodge ME, Tang W, Lu J, Ma Z, Fan CW, Wei S, Hao W, Kilgore J, Williams NS, Roth MG, Amatruda JF, Chen C, Lum L (2009) Small molecule-mediated disruption of Wnt-dependent signaling in tissue regeneration and cancer. Nat Chem Biol 5(2):100–107. https://doi.org/10.1038/nchembio.137

36. Ren Y, Lee MY, Schliffke S, Paavola J, Amos PJ, Ge X, Ye M, Zhu S, Senyei G, Lum L, Ehrlich BE, Qyang Y (2011) Small molecule Wnt inhibitors enhance the efficiency of BMP-4-directed cardiac differentiation of human pluripotent stem cells. J Mol Cell Cardiol 51(3):280–287. https://doi.org/10.1016/j.yjmcc.2011.04.012

37. Hudson J, Titmarsh D, Hidalgo A, Wolvetang E, Cooper-White J (2012) Primitive cardiac cells from human embryonic stem cells. Stem Cells Dev 21(9):1513–1523. https://doi.org/10.1089/scd.2011.0254

38. Lanier M, Schade D, Willems E, Tsuda M, Spiering S, Kalisiak J, Mercola M, Cashman JR (2012) Wnt inhibition correlates with human embryonic stem cell cardiomyogenesis: a structure-activity relationship study based on inhibitors for the Wnt response. J Med Chem 55(2):697–708. https://doi.org/10.1021/jm2010223

39. Breckwoldt K, Letuffe-Breniere D, Mannhardt I, Schulze T, Ulmer B, Werner T, Benzin A, Klampe B, Reinsch MC, Laufer S, Shibamiya A, Prondzynski M, Mearini G, Schade D, Fuchs S, Neuber C, Kramer E, Saleem U, Schulze ML, Rodriguez ML, Eschenhagen T,

Hansen A (2017) Differentiation of cardiomyocytes and generation of human engineered heart tissue. Nat Protoc 12(6):1177–1197. https://doi.org/10.1038/nprot.2017.033

40. Huang SM, Mishina YM, Liu S, Cheung A, Stegmeier F, Michaud GA, Charlat O, Wiellette E, Zhang Y, Wiessner S, Hild M, Shi X, Wilson CJ, Mickanin C, Myer V, Fazal A, Tomlinson R, Serluca F, Shao W, Cheng H, Shultz M, Rau C, Schirle M, Schlegl J, Ghidelli S, Fawell S, Lu C, Curtis D, Kirschner MW, Lengauer C, Finan PM, Tallarico JA, Bouwmeester T, Porter JA, Bauer A, Cong F (2009) Tankyrase inhibition stabilizes axin and antagonizes Wnt signalling. Nature 461(7264):614–620. https://doi.org/10.1038/nature08356

41. Wang H, Hao J, Hong CC (2011) Cardiac induction of embryonic stem cells by a small molecule inhibitor of Wnt/beta-catenin signaling. ACS Chem Biol 6(2):192–197. https://doi.org/10.1021/cb100323z

42. Minami I, Yamada K, Otsuji TG, Yamamoto T, Shen Y, Otsuka S, Kadota S, Morone N, Barve M, Asai Y, Tenkova-Heuser T, Heuser JE, Uesugi M, Aiba K, Nakatsuji N (2012) A small molecule that promotes cardiac differentiation of human pluripotent stem cells under defined, cytokine- and xeno-free conditions. Cell Rep 2(5):1448–1460. https://doi.org/10.1016/j.celrep.2012.09.015

43. Proffitt KD, Madan B, Ke Z, Pendharkar V, Ding L, Lee MA, Hannoush RN, Virshup DM (2013) Pharmacological inhibition of the Wnt acyltransferase PORCN prevents growth of WNT-driven mammary cancer. Cancer Res 73(2):502–507. https://doi.org/10.1158/0008-5472.can-12-2258

44. Burridge PW, Matsa E, Shukla P, Lin ZC, Churko JM, Ebert AD, Lan F, Diecke S, Huber B, Mordwinkin NM, Plews JR, Abilez OJ, Cui B, Gold JD, Wu JC (2014) Chemically defined generation of human cardiomyocytes. Nat Methods 11(8):855–860. https://doi.org/10.1038/nmeth.2999

45. Inman GJ, Nicolas FJ, Callahan JF, Harling JD, Gaster LM, Reith AD, Laping NJ, Hill CS (2002) SB-431542 is a potent and specific inhibitor of transforming growth factor-beta superfamily type I activin receptor-like kinase (ALK) receptors ALK4, ALK5, and ALK7. Mol Pharmacol 62(1):65–74

46. Kattman SJ, Witty AD, Gagliardi M, Dubois NC, Niapour M, Hotta A, Ellis J, Keller G (2011) Stage-specific optimization of activin/nodal and BMP signaling promotes cardiac differentiation of mouse and human pluripotent stem cell lines. Cell Stem Cell 8(2):228–240. https://doi.org/10.1016/j.stem.2010.12.008

47. Tojo M, Hamashima Y, Hanyu A, Kajimoto T, Saitoh M, Miyazono K, Node M, Imamura T (2005) The ALK-5 inhibitor A-83-01 inhibits Smad signaling and epithelial-to-mesenchymal transition by transforming growth factor-beta. Cancer Sci 96(11):791–800. https://doi.org/10.1111/j.1349-7006.2005.00103.x

48. Chen WP, Wu SM (2012) Small molecule regulators of postnatal Nkx2.5 cardiomyoblast proliferation and differentiation. J Cell Mol Med 16(5):961–965. https://doi.org/10.1111/j.1582-4934.2011.01513.x

49. Yu PB, Hong CC, Sachidanandan C, Babitt JL, Deng DY, Hoyng SA, Lin HY, Bloch KD, Peterson RT (2008) Dorsomorphin inhibits BMP signals required for embryogenesis and iron metabolism. Nat Chem Biol 4(1):33–41. https://doi.org/10.1038/nchembio.2007.54

50. Hao J, Ho JN, Lewis JA, Karim KA, Daniels RN, Gentry PR, Hopkins CR, Lindsley CW, Hong CC (2010) In vivo structure-activity relationship study of dorsomorphin analogues identifies selective VEGF and BMP inhibitors. ACS Chem Biol 5(2):245–253. https://doi.org/10.1021/cb9002865

51. Aguilar JS, Begum AN, Alvarez J, Zhang XB, Hong Y, Hao J (2015) Directed cardiomyogenesis of human pluripotent stem cells by modulating Wnt/beta-catenin and BMP signalling with small molecules. Biochem J 469(2):235–241. https://doi.org/10.1042/BJ20150186

52. Cuny GD, Yu PB, Laha JK, Xing X, Liu JF, Lai CS, Deng DY, Sachidanandan C, Bloch KD, Peterson RT (2008) Structure-activity relationship study of bone morphogenetic protein

(BMP) signaling inhibitors. Bioorg Med Chem Lett 18(15):4388–4392. https://doi.org/10.1016/j.bmcl.2008.06.052

53. Loh KM, Ang LT, Zhang J, Kumar V, Ang J, Auyeong JQ, Lee KL, Choo SH, Lim CY, Nichane M, Tan J, Noghabi MS, Azzola L, Ng ES, Durruthy-Durruthy J, Sebastiano V, Poellinger L, Elefanty AG, Stanley EG, Chen Q, Prabhakar S, Weissman IL, Lim B (2014) Efficient endoderm induction from human pluripotent stem cells by logically directing signals controlling lineage bifurcations. Cell Stem Cell 14(2):237–252. https://doi.org/10.1016/j.stem.2013.12.007

54. Uehata M, Ishizaki T, Satoh H, Ono T, Kawahara T, Morishita T, Tamakawa H, Yamagami K, Inui J, Maekawa M, Narumiya S (1997) Calcium sensitization of smooth muscle mediated by a rho-associated protein kinase in hypertension. Nature 389 (6654):990–994. https://doi.org/10.1038/40187

55. Fonoudi H, Ansari H, Abbasalizadeh S, Larijani MR, Kiani S, Hashemizadeh S, Zarchi AS, Bosman A, Blue GM, Pahlavan S, Perry M, Orr Y, Mayorchak Y, Vandenberg J, Talkhabi M, Winlaw DS, Harvey RP, Aghdami N, Baharvand H (2015) A universal and robust integrated platform for the scalable production of human cardiomyocytes from pluripotent stem cells. Stem Cells Transl Med 4(12):1482–1494. https://doi.org/10.5966/sctm.2014-0275

56. Vlahos CJ, Matter WF, Hui KY, Brown RF (1994) A specific inhibitor of phosphatidylinositol 3-kinase, 2-(4-morpholinyl)-8-phenyl-4H-1-benzopyran-4-one (LY294002). J Biol Chem 269 (7):5241–5248

57. Naito AT, Akazawa H, Takano H, Minamino T, Nagai T, Aburatani H, Komuro I (2005) Phosphatidylinositol 3-kinase-Akt pathway plays a critical role in early cardiomyogenesis by regulating canonical Wnt signaling. Circ Res 97(2):144–151. https://doi.org/10.1161/01.RES.0000175241.92285.f8

58. McLean AB, D'Amour KA, Jones KL, Krishnamoorthy M, Kulik MJ, Reynolds DM, Sheppard AM, Liu H, Xu Y, Baetge EE, Dalton S (2007) Activin A efficiently specifies definitive endoderm from human embryonic stem cells only when phosphatidylinositol 3-kinase signaling is suppressed. Stem Cells 25(1):29–38. https://doi.org/10.1634/stemcells.2006-0219

59. Barrett S, Biwersi C, Kaufman M, Tecle H, Warmus J (2002) Oxygenated esters of 4-iodo phenylamino benzhydroxamic acid. US Patent WO2002006213 A2. https://doi.org/10.1016/j.bmcl.2008.10.054

60. Bansal R, Magge S, Winkler S (2003) Specific inhibitor of FGF receptor signaling: FGF-2-mediated effects on proliferation, differentiation, and MAPK activation are inhibited by PD173074 in oligodendrocyte-lineage cells. J Neurosci Res 74(4):486–493. https://doi.org/10.1002/jnr.10773

61. Cuenda A, Rouse J, Doza YN, Meier R, Cohen P, Gallagher TF, Young PR, Lee JC (1995) SB 203580 is a specific inhibitor of a MAP kinase homologue which is stimulated by cellular stresses and interleukin-1. FEBS Lett 364(2):229–233. doi:0014-5793(95)00357-F [pii]

62. Graichen R, Xu X, Braam SR, Balakrishnan T, Norfiza S, Sieh S, Soo SY, Tham SC, Mummery C, Colman A, Zweigerdt R, Davidson BP (2008) Enhanced cardiomyogenesis of human embryonic stem cells by a small molecular inhibitor of p38 MAPK. Differentiation 76(4):357–370. https://doi.org/10.1111/j.1432-0436.2007.00236.x

63. Chen JK, Taipale J, Young KE, Maiti T, Beachy PA (2002) Small molecule modulation of Smoothened activity. Proc Natl Acad Sci U S A 99(22):14071–14076. https://doi.org/10.1073/pnas.182542899

64. Wu X, Ding S, Ding Q, Gray NS, Schultz PG (2002) A small molecule with osteogenesis-inducing activity in multipotent mesenchymal progenitor cells. J Am Chem Soc 124 (49):14520–14521

65. Borowiak M, Maehr R, Chen S, Chen AE, Tang W, Fox JL, Schreiber SL, Melton DA (2009) Small molecules efficiently direct endodermal differentiation of mouse and human embryonic stem cells. Cell Stem Cell 4(4):348–358. https://doi.org/10.1016/j.stem.2009.01.014

66. Levitzki A, Gazit A (1995) Tyrosine kinase inhibition: an approach to drug development. Science 267(5205):1782–1788

67. Zhu WZ, Xie Y, Moyes KW, Gold JD, Askari B, Laflamme MA (2010) Neuregulin/ErbB signaling regulates cardiac subtype specification in differentiating human embryonic stem cells. Circ Res 107(6):776–786. https://doi.org/10.1161/circresaha.110.223917

68. Devor DC, Singh AK, Frizzell RA, Bridges RJ (1996) Modulation of Cl− secretion by benzimidazolones. I. Direct activation of a Ca(2+)-dependent K+ channel. Am J Phys 271 (5 Pt 1):L775–L784

69. Jara-Avaca M, Kempf H, Ruckert M, Robles-Diaz D, Franke A, de la Roche J, Fischer M, Malan D, Sasse P, Solodenko W, Drager G, Kirschning A, Martin U, Zweigerdt R (2017) EBIO does not induce cardiomyogenesis in human pluripotent stem cells but modulates cardiac subtype enrichment by lineage-selective survival. Stem Cell Rep 8(2):305–317. https://doi.org/10.1016/j.stemcr.2016.12.012

70. Santilli AA, Scotese AC, Tomarelli RM (1974) A potent antihypercholesterolemic agent: (4-chloro-6-(2,3-xylidino)-2-pyrimidinylthio) acetic acid (Wy-14643). Experientia 30 (10):1110–1111

71. Poon E, Keung W, Liang Y, Ramalingam R, Yan B, Zhang S, Chopra A, Moore J, Herren A, Lieu DK, Wong HS, Weng Z, Wong OT, Lam YW, Tomaselli GF, Chen C, Boheler KR, Li RA (2015) Proteomic analysis of human pluripotent stem cell-derived, fetal, and adult ventricular cardiomyocytes reveals pathways crucial for cardiac metabolism and maturation. Circ Cardiovasc Genet 8(3):427–436. https://doi.org/10.1161/CIRCGENETICS.114.000918

72. Willems E, Spiering S, Davidovics H, Lanier M, Xia Z, Dawson M, Cashman J, Mercola M (2011) Small-molecule inhibitors of the Wnt pathway potently promote cardiomyocytes from human embryonic stem cell-derived mesoderm. Circ Res 109(4):360–364. https://doi.org/10.1161/CIRCRESAHA.111.249540

73. Lian X, Zhang J, Azarin SM, Zhu K, Hazeltine LB, Bao X, Hsiao C, Kamp TJ, Palecek SP (2013) Directed cardiomyocyte differentiation from human pluripotent stem cells by modulating Wnt/beta-catenin signaling under fully defined conditions. Nat Protoc 8(1):162–175. https://doi.org/10.1038/nprot.2012.150

74. Kempf H, Lecina M, Ting S, Zweigerdt R, Oh S (2011) Distinct regulation of mitogen-activated protein kinase activities is coupled with enhanced cardiac differentiation of human embryonic stem cells. Stem Cell Res 7(3):198–209. https://doi.org/10.1016/j.scr.2011.06.001

75. Kleger A, Seufferlein T, Malan D, Tischendorf M, Storch A, Wolheim A, Latz S, Protze S, Porzner M, Proepper C, Brunner C, Katz SF, Varma Pusapati G, Bullinger L, Franz WM, Koehntop R, Giehl K, Spyrantis A, Wittekindt O, Lin Q, Zenke M, Fleischmann BK, Wartenberg M, Wobus AM, Boeckers TM, Liebau S (2010) Modulation of calcium-activated potassium channels induces cardiogenesis of pluripotent stem cells and enrichment of pacemaker-like cells. Circulation 122(18):1823–1836. https://doi.org/10.1161/CIRCULATIONAHA.110.971721

76. Van Camp JK, Beckers S, Zegers D, Van Hul W (2014) Wnt signaling and the control of human stem cell fate. Stem Cell Rev 10(2):207–229. https://doi.org/10.1007/s12015-013-9486-8

77. Nusse R (2012) Wnt signaling. Cold Spring Harbor Perspect Biol 4(5):a011163. https://doi.org/10.1101/cshperspect.a011163

78. Valenta T, Hausmann G, Basler K (2012) The many faces and functions of beta-catenin. EMBO J 31(12):2714–2736. https://doi.org/10.1038/emboj.2012.150

79. Nusse R (2017) The Wnt homepage: Wnt target genes. http://web.stanford.edu/group/nusselab/cgi-bin/wnt/target_genes. Accessed 11 May 2017

80. Fang D, Hawke D, Zheng Y, Xia Y, Meisenhelder J, Nika H, Mills GB, Kobayashi R, Hunter T, Lu Z (2007) Phosphorylation of beta-catenin by AKT promotes beta-catenin transcriptional activity. J Biol Chem 282(15):11221–11229. https://doi.org/10.1074/jbc.M611871200

81. Hino S, Tanji C, Nakayama KI, Kikuchi A (2005) Phosphorylation of beta-catenin by cyclic AMP-dependent protein kinase stabilizes beta-catenin through inhibition of its ubiquitination. Mol Cell Biol 25(20):9063–9072. https://doi.org/10.1128/MCB.25.20.9063-9072.2005

82. Taurin S, Sandbo N, Qin Y, Browning D, Dulin NO (2006) Phosphorylation of beta-catenin by cyclic AMP-dependent protein kinase. J Biol Chem 281(15):9971–9976. https://doi.org/10.1074/jbc.M508778200

83. Tago K, Nakamura T, Nishita M, Hyodo J, Nagai S, Murata Y, Adachi S, Ohwada S, Morishita Y, Shibuya H, Akiyama T (2000) Inhibition of Wnt signaling by ICAT, a novel beta-catenin-interacting protein. Genes Dev 14(14):1741–1749

84. Ying QL, Wray J, Nichols J, Batlle-Morera L, Doble B, Woodgett J, Cohen P, Smith A (2008) The ground state of embryonic stem cell self-renewal. Nature 453(7194):519–523. https://doi.org/10.1038/nature06968

85. Davidson KC, Adams AM, Goodson JM, McDonald CE, Potter JC, Berndt JD, Biechele TL, Taylor RJ, Moon RT (2012) Wnt/beta-catenin signaling promotes differentiation, not self-renewal, of human embryonic stem cells and is repressed by Oct4. Proc Natl Acad Sci U S A 109(12):4485–4490. https://doi.org/10.1073/pnas.1118777109

86. Siller R, Greenhough S, Naumovska E, Sullivan GJ (2015) Small-molecule-driven hepatocyte differentiation of human pluripotent stem cells. Stem Cell Rep 4(5):939–952. https://doi.org/10.1016/j.stemcr.2015.04.001

87. Tan JY, Sriram G, Rufaihah AJ, Neoh KG, Cao T (2013) Efficient derivation of lateral plate and paraxial mesoderm subtypes from human embryonic stem cells through GSKi-mediated differentiation. Stem Cells Dev 22(13):1893–1906. https://doi.org/10.1089/scd.2012.0590

88. Araoka T, Mae S, Kurose Y, Uesugi M, Ohta A, Yamanaka S, Osafune K (2014) Efficient and rapid induction of human iPSCs/ESCs into nephrogenic intermediate mesoderm using small molecule-based differentiation methods. PLoS One 9(1):e84881. https://doi.org/10.1371/journal.pone.0084881

89. Lam AQ, Freedman BS, Morizane R, Lerou PH, Valerius MT, Bonventre JV (2014) Rapid and efficient differentiation of human pluripotent stem cells into intermediate mesoderm that forms tubules expressing kidney proximal tubular markers. J Am Soc Nephrol 25(6):1211–1225. https://doi.org/10.1681/ASN.2013080831

90. Shelton M, Metz J, Liu J, Carpenedo RL, Demers SP, Stanford WL, Skerjanc IS (2014) Derivation and expansion of PAX7-positive muscle progenitors from human and mouse embryonic stem cells. Stem Cell Rep 3(3):516–529. https://doi.org/10.1016/j.stemcr.2014.07.001

91. Voronkov A, Krauss S (2013) Wnt/beta-catenin signaling and small molecule inhibitors. Curr Pharm Des 19(4):634–664

92. Zweigerdt R (2007) The art of cobbling a running pump—will human embryonic stem cells mend broken hearts? Semin Cell Dev Biol 18(6):794–804. https://doi.org/10.1016/j.semcdb.2007.09.014

93. Zweigerdt R (2009) Large scale production of stem cells and their derivatives. Adv Biochem Eng Biotechnol 114:201–235. doi:10.1007/10_2008_27

94. Martin U (2015) New muscle for old hearts: engineering tissue from pluripotent stem cells. Hum Gene Ther 26(5):305–311. https://doi.org/10.1089/hum.2015.022

95. Taylor CJ, Bolton EM, Bradley JA (2011) Immunological considerations for embryonic and induced pluripotent stem cell banking. Philos Trans R Soc Lond Ser B Biol Sci 366(1575):2312–2322. https://doi.org/10.1098/rstb.2011.0030

96. Menasche P, Vanneaux V, Hagege A, Bel A, Cholley B, Cacciapuoti I, Parouchev A, Benhamouda N, Tachdjian G, Tosca L, Trouvin JH, Fabreguettes JR, Bellamy V, Guillemain R, Suberbielle Boissel C, Tartour E, Desnos M, Larghero J (2015) Human embryonic stem cell-derived cardiac progenitors for severe heart failure treatment: first clinical case report. Eur Heart J 36(30):2011–2017. https://doi.org/10.1093/eurheartj/ehv189

97. Kempf H, Andree B, Zweigerdt R (2016) Large-scale production of human pluripotent stem cell derived cardiomyocytes. Adv Drug Deliv Rev 96:18–30. https://doi.org/10.1016/j.addr. 2015.11.016
98. Kropp C, Massai D, Zweigerdt R (2016) Progress and challenges in large-scale expansion of human pluripotent stem cells. Process Biochem. https://doi.org/10.1016/j.procbio.2016.09. 032
99. Evans MJ, Kaufman MH (1981) Establishment in culture of pluripotential cells from mouse embryos. Nature 292(5819):154–156
100. Zweigerdt R, Burg M, Willbold E, Abts H, Ruediger M (2003) Generation of confluent cardiomyocyte monolayers derived from embryonic stem cells in suspension: a cell source for new therapies and screening strategies. Cytotherapy 5(5):399–413. https://doi.org/10. 1080/14653240310003062
101. Zandstra PW, Bauwens C, Yin T, Liu Q, Schiller H, Zweigerdt R, Pasumarthi KB, Field LJ (2003) Scalable production of embryonic stem cell-derived cardiomyocytes. Tissue Eng 9 (4):767–778. https://doi.org/10.1089/107632703768247449
102. Schroeder M, Niebruegge S, Werner A, Willbold E, Burg M, Ruediger M, Field LJ, Lehmann J, Zweigerdt R (2005) Differentiation and lineage selection of mouse embryonic stem cells in a stirred bench scale bioreactor with automated process control. Biotechnol Bioeng 92(7):920–933. https://doi.org/10.1002/bit.20668
103. Niebruegge S, Nehring A, Bar H, Schroeder M, Zweigerdt R, Lehmann J (2008) Cardiomyocyte production in mass suspension culture: embryonic stem cells as a source for great amounts of functional cardiomyocytes. Tissue Eng A 14(10):1591–1601. https://doi. org/10.1089/ten.tea.2007.0247
104. Dahlmann J, Kensah G, Kempf H, Skvorc D, Gawol A, Elliott DA, Drager G, Zweigerdt R, Martin U, Gruh I (2013) The use of agarose microwells for scalable embryoid body formation and cardiac differentiation of human and murine pluripotent stem cells. Biomaterials 34 (10):2463–2471. https://doi.org/10.1016/j.biomaterials.2012.12.024
105. Olmer R, Haase A, Merkert S, Cui W, Palecek J, Ran C, Kirschning A, Scheper T, Glage S, Miller K, Curnow EC, Hayes ES, Martin U (2010) Long term expansion of undifferentiated human iPS and ES cells in suspension culture using a defined medium. Stem Cell Res 5 (1):51–64. https://doi.org/10.1016/j.scr.2010.03.005
106. Kehat I, Kenyagin-Karsenti D, Snir M, Segev H, Amit M, Gepstein A, Livne E, Binah O, Itskovitz-Eldor J, Gepstein L (2001) Human embryonic stem cells can differentiate into myocytes with structural and functional properties of cardiomyocytes. J Clin Invest 108 (3):407–414. https://doi.org/10.1172/JCI12131
107. Xu XQ, Soo SY, Sun W, Zweigerdt R (2009) Global expression profile of highly enriched cardiomyocytes derived from human embryonic stem cells. Stem Cells 27(9):2163–2174. https://doi.org/10.1002/stem.166
108. Xu XQ, Zweigerdt R, Soo SY, Ngoh ZX, Tham SC, Wang ST, Graichen R, Davidson B, Colman A, Sun W (2008) Highly enriched cardiomyocytes from human embryonic stem cells. Cytotherapy 10(4):376–389. https://doi.org/10.1080/14653240802105307
109. Olmer R, Lange A, Selzer S, Kasper C, Haverich A, Martin U, Zweigerdt R (2012) Suspension culture of human pluripotent stem cells in controlled, stirred bioreactors. Tissue Eng Part C Methods 18(10):772–784. https://doi.org/10.1089/ten.TEC.2011.0717
110. Zweigerdt R, Olmer R, Singh H, Haverich A, Martin U (2011) Scalable expansion of human pluripotent stem cells in suspension culture. Nat Protoc 6(5):689–700. https://doi.org/10. 1038/nprot.2011.318
111. Ting S, Lecina M, Chan YC, Tse HF, Reuveny S, Oh SK (2013) Nutrient supplemented serum-free medium increases cardiomyogenesis efficiency of human pluripotent stem cells. World J Stem Cells 5(3):86–97. https://doi.org/10.4252/wjsc.v5.i3.86
112. Kensah G, Roa Lara A, Dahlmann J, Zweigerdt R, Schwanke K, Hegermann J, Skvorc D, Gawol A, Azizian A, Wagner S, Maier LS, Krause A, Drager G, Ochs M, Haverich A, Gruh I, Martin U (2013) Murine and human pluripotent stem cell-derived cardiac bodies form

contractile myocardial tissue in vitro. Eur Heart J 34(15):1134–1146. https://doi.org/10.1093/eurheartj/ehs349

113. Ueno S, Weidinger G, Osugi T, Kohn AD, Golob JL, Pabon L, Reinecke H, Moon RT, Murry CE (2007) Biphasic role for Wnt/beta-catenin signaling in cardiac specification in zebrafish and embryonic stem cells. Proc Natl Acad Sci U S A 104(23):9685–9690. https://doi.org/10.1073/pnas.0702859104

114. Kempf H, Kropp C, Olmer R, Martin U, Zweigerdt R (2015) Cardiac differentiation of human pluripotent stem cells in scalable suspension culture. Nat Protoc 10(9):1345–1361. https://doi.org/10.1038/nprot.2015.089

115. Kempf H, Olmer R, Kropp C, Ruckert M, Jara-Avaca M, Robles-Diaz D, Franke A, Elliott DA, Wojciechowski D, Fischer M, Roa Lara A, Kensah G, Gruh I, Haverich A, Martin U, Zweigerdt R (2014) Controlling expansion and cardiomyogenic differentiation of human pluripotent stem cells in scalable suspension culture. Stem Cell Rep 3(6):1132–1146. https://doi.org/10.1016/j.stemcr.2014.09.017

116. Kropp C, Kempf H, Halloin C, Robles-Diaz D, Franke A, Scheper T, Kinast K, Knorpp T, Joos TO, Haverich A, Martin U, Zweigerdt R, Olmer R (2016) Impact of feeding strategies on the scalable expansion of human pluripotent stem cells in single-use stirred tank bioreactors. Stem Cells Transl Med 5(10):1289–1301. https://doi.org/10.5966/sctm.2015-0253

117. Zhang M, Schulte JS, Heinick A, Piccini I, Rao J, Quaranta R, Zeuschner D, Malan D, Kim KP, Ropke A, Sasse P, Arauzo-Bravo M, Seebohm G, Scholer H, Fabritz L, Kirchhof P, Muller FU, Greber B (2015) Universal cardiac induction of human pluripotent stem cells in two and three-dimensional formats: implications for in vitro maturation. Stem Cells 33 (5):1456–1469. https://doi.org/10.1002/stem.1964

118. Lian X, Bao X, Zilberter M, Westman M, Fisahn A, Hsiao C, Hazeltine LB, Dunn KK, Kamp TJ, Palecek SP (2015) Chemically defined, albumin-free human cardiomyocyte generation. Nat Methods 12(7):595–596. https://doi.org/10.1038/nmeth.3448

119. Crook JM, Peura TT, Kravets L, Bosman AG, Buzzard JJ, Horne R, Hentze H, Dunn NR, Zweigerdt R, Chua F, Upshall A, Colman A (2007) The generation of six clinical-grade human embryonic stem cell lines. Cell Stem Cell 1(5):490–494

120. Haase A, Olmer R, Schwanke K, Wunderlich S, Merkert S, Hess C, Zweigerdt R, Gruh I, Meyer J, Wagner S, Maier LS, Han DW, Glage S, Miller K, Fischer P, Scholer HR, Martin U (2009) Generation of induced pluripotent stem cells from human cord blood. Cell Stem Cell 5 (4):434–441. https://doi.org/10.1016/j.stem.2009.08.021

121. Chen VC, Ye J, Shukla P, Hua G, Chen D, Lin Z, Liu JC, Chai J, Gold J, Wu J, Hsu D, Couture LA (2015) Development of a scalable suspension culture for cardiac differentiation from human pluripotent stem cells. Stem Cell Res 15(2):365–375. https://doi.org/10.1016/j.scr.2015.08.002

122. Chong JJ, Yang X, Don CW, Minami E, Liu YW, Weyers JJ, Mahoney WM, Van Biber B, Cook SM, Palpant NJ, Gantz JA, Fugate JA, Muskheli V, Gough GM, Vogel KW, Astley CA, Hotchkiss CE, Baldessari A, Pabon L, Reinecke H, Gill EA, Nelson V, Kiem HP, Laflamme MA, Murry CE (2014) Human embryonic-stem-cell-derived cardiomyocytes regenerate non-human primate hearts. Nature 510(7504):273–277. https://doi.org/10.1038/nature13233

123. Mauritz C, Martens A, Rojas SV, Schnick T, Rathert C, Schecker N, Menke S, Glage S, Zweigerdt R, Haverich A, Martin U, Kutschka I (2011) Induced pluripotent stem cell (iPSC)-derived Flk-1 progenitor cells engraft, differentiate, and improve heart function in a mouse model of acute myocardial infarction. Eur Heart J 32(21):2634–2641. https://doi.org/10.1093/eurheartj/ehr166

124. Shen N, Knopf A, Westendorf C, Kraushaar U, Riedl J, Bauer H, Poschel S, Layland SL, Holeiter M, Knolle S, Brauchle E, Nsair A, Hinderer S, Schenke-Layland K (2017) Steps toward maturation of embryonic stem cell-derived cardiomyocytes by defined physical signals. Stem Cell Rep 9(1):122–135. https://doi.org/10.1016/j.stemcr.2017.04.021

125. Van den Berg CW, Okawa S, Chuva de Sousa Lopes SM, van Iperen L, Passier R, Braam SR, Tertoolen LG, del Sol A, Davis RP, Mummery CL (2015) Transcriptome of human foetal

heart compared with cardiomyocytes from pluripotent stem cells. Development 142 (18):3231–3238. https://doi.org/10.1242/dev.123810

126. Nunes SS, Miklas JW, Liu J, Aschar-Sobbi R, Xiao Y, Zhang B, Jiang J, Masse S, Gagliardi M, Hsieh A, Thavandiran N, Laflamme MA, Nanthakumar K, Gross GJ, Backx PH, Keller G, Radisic M (2013) Biowire: a platform for maturation of human pluripotent stem cell-derived cardiomyocytes. Nat Methods 10(8):781–787. https://doi.org/10.1038/nmeth.2524

127. Tiburcy M, Hudson JE, Balfanz P, Schlick S, Meyer T, Chang Liao ML, Levent E, Raad F, Zeidler S, Wingender E, Riegler J, Wang M, Gold JD, Kehat I, Wettwer E, Ravens U, Dierickx P, van Laake LW, Goumans MJ, Khadjeh S, Toischer K, Hasenfuss G, Couture LA, Unger A, Linke WA, Araki T, Neel B, Keller G, Gepstein L, Wu JC, Zimmermann WH (2017) Defined engineered human myocardium with advanced maturation for applications in heart failure modeling and repair. Circulation 135(19):1832–1847. https://doi.org/10.1161/CIRCULATIONAHA.116.024145

128. Andree B, Bar A, Haverich A, Hilfiker A (2013) Small intestinal submucosa segments as matrix for tissue engineering: review. Tissue Eng Part B Rev 19(4):279–291. https://doi.org/10.1089/ten.TEB.2012.0583

129. Weinberger F, Mannhardt I, Eschenhagen T (2017) Engineering cardiac muscle tissue: a maturating field of research. Circ Res 120(9):1487–1500. https://doi.org/10.1161/CIRCRESAHA.117.310738

130. Weber N, Schwanke K, Greten S, Wendland M, Iorga B, Fischer M, Geers-Knorr C, Hegermann J, Wrede C, Fiedler J, Kempf H, Franke A, Piep B, Pfanne A, Thum T, Martin U, Brenner B, Zweigerdt R, Kraft T (2016) Stiff matrix induces switch to pure beta-cardiac myosin heavy chain expression in human ESC-derived cardiomyocytes. Basic Res Cardiol 111(6):68. https://doi.org/10.1007/s00395-016-0587-9

131. Lundy SD, Zhu WZ, Regnier M, Laflamme MA (2013) Structural and functional maturation of cardiomyocytes derived from human pluripotent stem cells. Stem Cells Dev 22 (14):1991–2002. https://doi.org/10.1089/scd.2012.0490

132. Ribeiro AJ, Ang YS, Fu JD, Rivas RN, Mohamed TM, Higgs GC, Srivastava D, Pruitt BL (2015) Contractility of single cardiomyocytes differentiated from pluripotent stem cells depends on physiological shape and substrate stiffness. Proc Natl Acad Sci U S A 112 (41):12705–12710. https://doi.org/10.1073/pnas.1508073112

133. Suliman HB, Zobi F, Piantadosi CA (2016) Heme oxygenase-1/carbon monoxide system and embryonic stem cell differentiation and maturation into cardiomyocytes. Antioxid Redox Signal 24(7):345–360. https://doi.org/10.1089/ars.2015.6342

134. Kraft T, Witjas-Paalberends ER, Boontje NM, Tripathi S, Brandis A, Montag J, Hodgkinson JL, Francino A, Navarro-Lopez F, Brenner B, Stienen GJ, van der Velden J (2013) Familial hypertrophic cardiomyopathy: functional effects of myosin mutation R723G in cardiomyocytes. J Mol Cell Cardiol 57:13–22. https://doi.org/10.1016/j.yjmcc.2013.01.001

135. Tohyama S, Hattori F, Sano M, Hishiki T, Nagahata Y, Matsuura T, Hashimoto H, Suzuki T, Yamashita H, Satoh Y, Egashira T, Seki T, Muraoka N, Yamakawa H, Ohgino Y, Tanaka T, Yoichi M, Yuasa S, Murata M, Suematsu M, Fukuda K (2013) Distinct metabolic flow enables large-scale purification of mouse and human pluripotent stem cell-derived cardiomyocytes. Cell Stem Cell 12(1):127–137. https://doi.org/10.1016/j.stem.2012.09.013

136. Fuerstenau-Sharp M, Zimmermann ME, Stark K, Jentsch N, Klingenstein M, Drzymalski M, Wagner S, Maier LS, Hehr U, Baessler A, Fischer M, Hengstenberg C (2015) Generation of highly purified human cardiomyocytes from peripheral blood mononuclear cell-derived induced pluripotent stem cells. PLoS One 10(5):e0126596. https://doi.org/10.1371/journal.pone.0126596

137. Drawnel FM, Boccardo S, Prummer M, Delobel F, Graff A, Weber M, Gerard R, Badi L, Kam-Thong T, Bu L, Jiang X, Hoflack JC, Kiialainen A, Jeworutzki E, Aoyama N, Carlson C, Burcin M, Gromo G, Boehringer M, Stahlberg H, Hall BJ, Magnone MC, Kolaja K, Chien KR, Bailly J, Iacone R (2014) Disease modeling and phenotypic drug

screening for diabetic cardiomyopathy using human induced pluripotent stem cells. Cell Rep 9(3):810–821. https://doi.org/10.1016/j.celrep.2014.09.055

138. Yang X, Rodriguez M, Pabon L, Fischer KA, Reinecke H, Regnier M, Sniadecki NJ, Ruohola-Baker H, Murry CE (2014) Tri-iodo-L-thyronine promotes the maturation of human cardiomyocytes-derived from induced pluripotent stem cells. J Mol Cell Cardiol 72:296–304. https://doi.org/10.1016/j.yjmcc.2014.04.005

139. Clevers H (2016) Modeling development and disease with organoids. Cell 165 (7):1586–1597. https://doi.org/10.1016/j.cell.2016.05.082

140. Passier R, Orlova V, Mummery C (2016) Complex tissue and disease modeling using hiPSCs. Cell Stem Cell 18(3):309–321. https://doi.org/10.1016/j.stem.2016.02.011

141. Giacomelli E, Bellin M, Sala L, van Meer BJ, Tertoolen LG, Orlova VV, Mummery CL (2017) Three-dimensional cardiac microtissues composed of cardiomyocytes and endothelial cells co-differentiated from human pluripotent stem cells. Development 144(6):1008–1017. https://doi.org/10.1242/dev.143438

142. Ma Z, Wang J, Loskill P, Huebsch N, Koo S, Svedlund FL, Marks NC, Hua EW, Grigoropoulos CP, Conklin BR, Healy KE (2015) Self-organizing human cardiac microchambers mediated by geometric confinement. Nat Commun 6:7413. https://doi.org/10.1038/ncomms8413

143. Moretti A, Bellin M, Welling A, Jung CB, Lam JT, Bott-Flugel L, Dorn T, Goedel A, Hohnke C, Hofmann F, Seyfarth M, Sinnecker D, Schomig A, Laugwitz KL (2010) Patient-specific induced pluripotent stem-cell models for long-QT syndrome. N Engl J Med 363(15):1397–1409. https://doi.org/10.1056/NEJMoa0908679

144. Friedrichs S, Malan D, Sasse P (2013) Modeling long QT syndromes using induced pluripotent stem cells: current progress and future challenges. Trends Cardiovasc Med 23(4):91–98. https://doi.org/10.1016/j.tcm.2012.09.006

145. Burridge PW, Li YF, Matsa E, Wu H, Ong SG, Sharma A, Holmstrom A, Chang AC, Coronado MJ, Ebert AD, Knowles JW, Telli ML, Witteles RM, Blau HM, Bernstein D, Altman RB, Wu JC (2016) Human induced pluripotent stem cell-derived cardiomyocytes recapitulate the predilection of breast cancer patients to doxorubicin-induced cardiotoxicity. Nat Med 22(5):547–556. https://doi.org/10.1038/nm.4087

146. Rojas SV, Martens A, Zweigerdt R, Baraki H, Rathert C, Schecker N, Rojas-Hernandez S, Schwanke K, Martin U, Haverich A, Kutschka I (2015) Transplantation effectiveness of induced pluripotent stem cells is improved by a fibrinogen biomatrix in an experimental model of ischemic heart failure. Tissue Eng Part A 21(13–14):1991–2000. https://doi.org/10.1089/ten.TEA.2014.0537

147. Templin C, Zweigerdt R, Schwanke K, Olmer R, Ghadri JR, Emmert MY, Müller E, Küest SM, Cohrs S, Schibli R, Kronen R, Hilbe M, Reinisch A, Strunk D, Haverich A, Hoerstrup S, Lüscher TF, Kaufmann PA, Landmesser U, Martin U (2012) Transplantation and tracking of human induced pluripotent stem cells in a pig model of myocardial infarction: assessment of cell survival, engraftment and distribution by hybrid SPECT-CT imaging of sodium iodide Symporter Trangene expression. Circulation 126(4):430–439

148. van den Akker F, Feyen DA, van den Hoogen P, van Laake LW, van Eeuwijk EC, Hoefer I, Pasterkamp G, Chamuleau SA, Grundeman PF, Doevendans PA, Sluijter JP (2017) Intramyocardial stem cell injection: go(ne) with the flow. Eur Heart J 38(3):184–186. https://doi.org/10.1093/eurheartj/ehw056

Adv Biochem Eng Biotechnol (2018) 163: 71–116
DOI: 10.1007/10_2017_27
© Springer International Publishing AG 2017
Published online: 26 October 2017

Specific Cell (Re-)Programming: Approaches and Perspectives

Frauke Hausburg, Julia Jeannine Jung, and Robert David

Abstract Many disorders are manifested by dysfunction of key cell types or their disturbed integration in complex organs. Thereby, adult organ systems often bear restricted self-renewal potential and are incapable of achieving functional regeneration. This underlies the need for novel strategies in the field of cell (re-)programming-based regenerative medicine as well as for drug development in vitro. The regenerative field has been hampered by restricted availability of adult stem cells and the potentially hazardous features of pluripotent embryonic stem cells (ESCs) and induced pluripotent stem cells (iPSCs). Moreover, ethical concerns and legal restrictions regarding the generation and use of ESCs still exist. The establishment of direct reprogramming protocols for various therapeutically valuable somatic cell types has overcome some of these limitations. Meanwhile, new perspectives for safe and efficient generation of different specified somatic cell types have emerged from numerous approaches relying on exogenous expression of lineage-specific transcription factors, coding and noncoding RNAs, and chemical compounds.

It should be of highest priority to develop protocols for the production of mature and physiologically functional cells with properties ideally matching those of their endogenous counterparts. Their availability can bring together basic research, drug screening, safety testing, and ultimately clinical trials. Here, we highlight the remarkable successes in cellular (re-)programming, which have greatly advanced the field of regenerative medicine in recent years. In particular, we review recent progress on the generation of cardiomyocyte subtypes, with a focus on cardiac pacemaker cells.

F. Hausburg, J.J. Jung, and R. David (✉)
Reference and Translation Center for Cardiac Stem Cell Therapy (RTC), Department of Cardiac Surgery, Rostock University Medical Center, Schillingallee 69, 18057 Rostock, Germany

Department Life, Light and Matter of the Interdisciplinary Faculty at Rostock University, Albert-Einstein-Strasse 25, 18059 Rostock, Germany
e-mail: robert.david@med.uni-rostock.de

Graphical Abstract

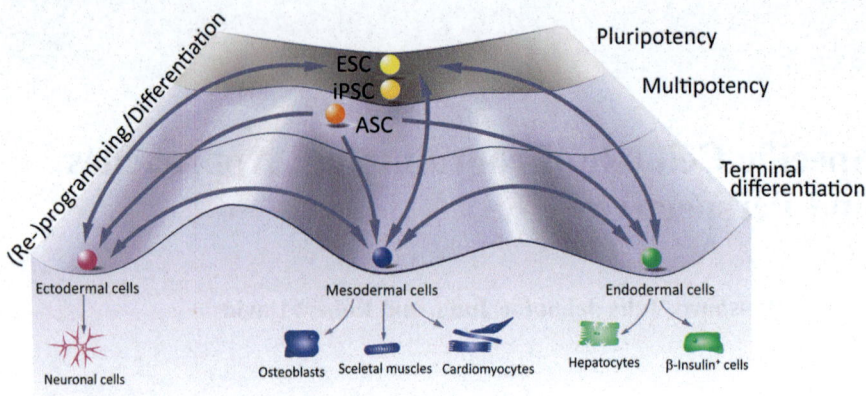

Keywords Cardiovascular regeneration, Cell fate conversion, Direct reprogramming, Lineage conversion, Metabolic disorders, Neurodegenerative disorders, Regenerative medicine

Contents

Abbreviations

(±)-BayK-8644	Ca^{2+} channel agonist
A83-01	TGF-β inhibitor
AA	Ascorbic acid
ACTN2	α-Actinin

ADSC	Adipose tissue-derived mesenchymal stem cell
AFP	α-Fetoprotein
Akt1	AKT serine/threonine kinase 1
ALB	Albumin
ALK5	TGFβ type I receptor kinase
ALP	Alkaline phosphatase
ANF	NPPA, natriuretic peptide A
APD	Action potential duration
APOA1	Apolipoprotein A1
AS8351	Iron chelator
ASC	Adult stem cell
Ascl1	Achaete-scute homolog 1
ATF5	Activating transcription factor 5
ATSC	Adipose tissue-derived mesenchymal stem cells
AV	Atrioventricular
Bcl2	B-cell lymphoma 2
BCT	Bioartificial cardiac tissue
bFGF	Basic fibroblast growth factor
bHLH	Basic helix-loop-helix
BIO	6-Bromoindirubin-3′-oxime, canonical Wnt activator
BIX01294	Diazepin-quinazolinamine derivative; histone-lysine methyltransferase inhibitor
Bmi1	BMI1 proto-oncogene, polycomb ring finger
BM-MSC	Bone marrow-derived mesenchymal stem cell
bpm	Beats per minute
Bry	Brachyury
CD166	ALCAM; activated leukocyte cell adhesion molecule
CEBPA	CCAAT/enhancer binding protein alpha
CF	Cardiac fibroblast
CHD	Congenital heart defect
CHIR	CHIR99021, GSK-3 inhibitor, Wnt activator
CM	Cardiomyocyte
C-MYC	MYC proto-oncogene, bHLH transcription factor
CPC	Cardiac progenitor cell
CRM	Cardiac reprogramming medium
CS	Conduction system
CT99021	SHH and the GSK3β inhibitor
cTnI	Troponin I3, cardiac type
cTnT	Troponin T2, cardiac type
Cx	Gap junction protein
CYP	Cytochrome P450
DAPT	N-[N-(3,5-Difluorophenacetyl)-L-alanyl]-S-phenylglycine t-butyl ester
DCX	Newborn neuron
DES	Desmin

DF	Dermal fibroblast
DFSC	Dental follicular-derived mesenchymal stem cell
DLX1	Distal-less homeobox
DMEM/F12	Dulbecco's modified eagle medium: nutrient mixture F-12
DMD	Dystrophin
EAD	Early after depolarizations
EBIO	1-EBIO; $K_{Ca}2/3$ channel activator
EGF	Epidermal growth factor
EPC	Endothelial progenitor cell
EPDC	Epicardium-derived cell
ESC	Embryonic stem cell
FFV	FGF2, FGF10 & VEGF
FGF	Fibroblast growth factor
FHF	First heart field
FLF	Fetal limb fibroblast
forskolin	Adenylyl cyclase activator
FOX	Forkhead box
GABA	Gamma-aminobutyric acid
GAD67	Glutamate decarboxylase 1
Gata4	GATA binding protein 4
GF	Gingival fibroblast
Glut2	SLC2A2; solute carrier family 2 member 2
GMT	Gata4, Mef2c & Tbx5
GMTH	Gata4, Mef2c, Tbx5 & Hand2
GO6983	PKC inhibitor
GSK126	Selective EZH2 methyltransferase inhibitor
Hand2	Heart and neural crest derivatives expressed 2
HC	Hepatocytes
hCMVEC	Human cardiac microvascular endothelial cell
HCN	Hyperpolarization-activated cyclic nucleotide channel
hEF	Human embryonic fibroblast
HFF	Human foreskin fibroblast
hiPSC-ECM	Induced pluripotent stem cell-derived embryonic cardiac myocyte
HNF	Hepatic nuclear factor
I-BET151	Bromodomain and extra-terminal domain family inhibitor
If	Funny current
iPSC	Induced pluripotent stem cell
Isl1	ISL LIM homeobox 1
ISX9	Neurogenesis inducer
ITS	Insulin-transferrin-selenium
JAK inhibitor I	Janus-Associated Kinase Inhibitor I
JNJ10198409	ATP-competitive inhibitor of platelet-derived growth Factor receptor tyrosine kinase
JNK	C-Jun N-terminal kinases

KLF4	Kruppel like factor 4
LDL	Low-density lipoprotein
LDN193189	BMP4 inhibitor
LF	Lung fibroblast
Lhx6	LIM homeobox protein 6
LIF	Leukemia inhibiting factor, JAK/STAT activator
LMX1A	LIM homeobox transcription factor 1 alpha
L-MYC	MYCL proto-oncogene, bHLH transcription factor
lncRNA	Long noncoding RNA
LVEF	Left ventricular ejection fraction
Ly294002	Phosphoinositide 3-kinase (PI3K) inhibitor, TGF-β activator
MafA	v-maf musculoaponeurotic fibrosarcoma oncogene family, protein A
Map2	Microtubule-associated protein 2
MAPK	Mitogen-activated protein kinase 1
MEF	Mouse embryonic fibroblast
Mef2c	Myocyte enhancer factor 2C
MF	Myofibroblast
Mhc	Myosin heavy chain
MI	Myocardial infarction
miR	microRNA
MLC2v	Myosin, light polypeptide 2, regulatory, cardiac, slow
MM3-GHT	Combination of Gata4, Hand2, Tbx5, and the fusion gene MM_3 between Mef2c and the transactivation domain of MyoD
MRI	Magnet resonance imaging
MSC	Mesenchymal stem cell
MYH3	Embryonic myosin
MYH6	Myosin heavy chain 6, cardiac muscle, alpha
MyHC	Myosin heavy chain 6
MyoD	Myogenic differentiation 1
MYOG	Myogenin
MYT1L	Myelin transcription factor 1 like
N2	Cysteine proteinase inhibitor
NeuN	Neuronal nuclei
NeuroD1	Neurogenic differentiation 1
NEUROD2	Neuronal differentiation 2
NFF	Neonatal foreskin fibroblast
NG2	Oligodendrocyte precursor
Ngn	Neurogenin
NKX	Homeobox protein
NMDA	N-Methyl-D-aspartate
NNCF	Neonatal cardiac fibroblast
NNF	Neonatal fibroblast
NRVM	Neonatal rat ventricular myocyte
NURR1	Nuclear receptor related 1 protein
OAC2	Oct4-activating compound 2

OB	Osteoblast
OC	Osteocyte
OCT4	POU class 5 homeobox 1
PC	Pacemaker cell
PD0325901	MEK1/2 inhibitor
Pdx1	Insulin promoter factor 1
pkc	Protein kinase C
PM	Pacemaker
PROX1	prospero homeobox 1
PSC	Pluripotent stem cell
Purmo	Purmorphamine
PV	Parvalbumin
Repsox	Inhibitor of the TGF-β type 1 receptor
ROCK	Rho-associated protein kinase
RUNX2	Runt related transcription factor 2
Ryr2	Ryanodine receptor 2
SAG	Smoothened agonist
SAN	Sinoatrial node
SB431542	TGF-β inhibitor
SC	Stem cell
SC1	Pluripotin, dual selective inhibitor of the ERK1 and Ras-GAP signaling pathways
SCD	Sudden cardiac death
SCN5A	Sodium channel, voltage-gated, type V, alpha subunit
SERPINA1	Serpin family A member 1
SHF	Second heart field
SHH	Sonic hedgehog
Shox2	Short stature homeobox 2
shRNA	Small hairpin RNA
siRNA	Small interfering RNA
SIRPA	Signal regulatory protein alpha
SLC1A2	Solute carrier family 1 member 2
Sox	Sex determining region Y-box 2
SP600125	JNK inhibitor
SR-3677	ROCK inhibitor
SSS	Sick sinus syndrome
STAT3	Signal transducer and activator of transcription 3
SU16F	Platelet-derived growth factor receptor β inhibitor
SV40	Simian vacuolating virus 40
Tbx	T-box factor 18
TF	Transcription factor
TGF-β	Transforming growth factor-β
THF	Tertiary heart field
Tnnt2	Troponin T2, cardiac type
TTF	Tail tip fibroblast

TTNPB	Analog of retinoic acid
TUBB3	β-III-tubulin
Tuj1	Neuron-specific class III beta-tubulin
Tzv	Thiazovivin
UNC0638	Histone methyltransferase inhibitor
VEGF	Vascular endothelial growth factor
VGLUT1	Vesicular glutamate transporter 1
VPA	Valporic acid
XAV939	Wnt inhibitor
Y-27632	ROCK inhibitor
α-MHC	Myosin heavy chain 6
βMe	β-Mercaptoethanol

1 Introduction

Remarkable impulses in basic research have opened up new perspectives in the field of regenerative medicine for hitherto unsolvable problems in conventional medicine. Over the last few decades, numerous efforts have led to deeper understanding and increased awareness of the invaluable advantages of new therapeutic strategies. These novel approaches may provide solutions to the challenges accompanying an ageing society and its consequences for the healthcare system. Moreover, the demand for patient-specific therapies to ensure prolonged health and quality of life will increase.

Many diseases such as cancers and neurodegenerative (e.g., Alzheimer or Parkinson disease), cardiovascular (e.g., ischemic heart disease and stroke), and metabolic disorders (e.g., diabetes mellitus) are associated with cell dysfunctions or abnormal cell–cell interactions. However, due to the restricted regenerative potential of many adult organs, functional repair of the affected tissue is often impossible (e.g., for the human heart) [1, 2]. This becomes even more evident if bearing in mind that among the top ten causes of death globally, 15 million patients died from cardiovascular disease in 2015, 1.59 million from diabetes mellitus, and 1.54 million from Alzheimer disease and other dementias [3]. Furthermore, the only promising therapeutic option for patients with end-stage organ failure remains organ transplantation. Yet, a major limitation of this approach is the shortage of donor organs. In 2015, only 605 donor hearts (1,606 livers, 259 pancreata) were successfully transplanted in the Eurotransplant region while 1,170 heart recipients (1,835 liver, 418 pancreata) were on the active Eurotransplant waiting list (at year-end in 2016). Thereby, in 2015, 209 heart patients (478 liver, 28 pancreata) on this waiting list died before they could receive the required organ transplant [4]. This underpins the importance of appropriate alternatives to organ transplantation, and of systems for patient-specific disease modeling and drug development [5–7].

The early experiments of the 1950s explored the transplantation of single cells instead of a whole organ. The first successful transplantation of multipotent adult stem cells (bone-marrow derived) was performed in 1957 between identical twins, the recipient with leukemia. This groundbreaking therapeutic approach led to the 1990 Nobel Prize in Physiology or Medicine for E. Donnall Thomas together with Joseph E. Murray "for their discoveries concerning organ and cell transplantation in the treatment of human disease" (http://www.nobelprize.org/nobel_prizes/medi cine/laureates/1990/). Today, the concept has been extended to numerous clinical studies on bone marrow-derived mesenchymal stem cell (BM-MSC) therapy against cardiovascular disease, evaluating the effectiveness and aiming at translation from bench to bedside [8–12]. However, current results provide only modest therapeutic improvement [9]. Accordingly, transplantation of, for example, CD133$^+$ adult stem cells (ASCs) together with coronary artery bypass graft surgery led only to a marginal improvement of the left ventricular ejection fraction (LVEF) by about 6% after 6 months [13]. Similar results were achieved in patients with acute myocardial infarction, summarized in a meta-analysis by Wang et al., which confirmed no significant increase in LVEF (1.47% improvement) [14]. Comprehensive long-term studies with considerable numbers and cohorts of patients are required to clarify issues regarding safety and efficiency.

In contrast, pluripotent stem cells (PSCs) such as embryonic stem cells (ESCs) and induced pluripotent stem cells (iPSCs) have not yet been completely examined scientifically and are not on the same clinical trial level as ASCs. However, they promise a much greater potential for medical innovation with regard to their self-renewal capacity and multilineage differentiation potential into all embryonic germ layers (endoderm, mesoderm, and ectoderm) and their derivatives [15–19]. Currently, advances in PSC research for clinical application are mainly in the fields of age-related macular degeneration, Parkinson disease, spinal cord injury, type I diabetes, and myocardial infarction. Phase I and II clinical trials to evaluate safety and therapeutic benefits are summarized by Trounson and DeWitt [19]. In June 2017, the Human Pluripotent Stem Cell registry (hPSCreg) listed 1,264 cell lines (hESC 706 and hiPSC 558; http://hpscreg.eu/). Worldwide, most of the hESCs are recorded in the USA and hiPSCs in the UK.

However, hESCs entail some serious disadvantages, primarily ethical and immunological concerns. For this reason, stem cell research is controlled by stringent legislation, but regulations differ within the European Union. Research in Belgium, Sweden, and the UK is allowed under certain conditions, whereas it is mostly prohibited in Lithuania, Poland, Germany, Slovakia, Austria, and Italy (http://hpscreg.eu/). The relatively new and exciting class of iPSCs could give deeper insights into developmental biology, thereby avoiding ethical concerns about hESCs. After murine and human iPSC lines were successfully generated, based on somatic reprogramming using the famous four Yamanaka factors [octamer-binding transcription factor 3/4 (Oct3/4), sex determining region Y-box 2 (Sox2), Krueppel-like factor 4 (Klf4), and c-Myc)] [20, 21], a number of concepts were established and offer great suitability for personalized disease modeling, drug development, and cell replacement therapies.

Because pluripotent cell behavior is associated with teratoma formation in vivo [22–25], it is advisable to explore further opportunities that avoid an intermediate pluripotent stage. Moreover, the application of human PSC-derived cell types is still limited by numerous hurdles, such as the time-consuming production of cardiomyocytes (CMs) [26] and the resulting high cost. Furthermore, the consequences of genetic and epigenetic alterations present in ESCs and iPSCs remain unclear [27]. The exciting approach of directly reprogramming one terminally differentiated somatic cell to another somatic cell of the same germ layer or even across germ layers is an alternative to PSC reprogramming. Such potential was shown the first time in 1987 with the cell fate conversion of murine fibroblasts into skeletal muscle cells using forced exogenous overexpression of the key transcription factor (TF) MyoD [28]. Nowadays, promising protocols propose the application of lineage-specific TFs, noncoding RNAs, key signaling pathway modulators, and functional substances to promote differentiation and maturation of various cell types. Moreover, a major advantage could be the use and differentiation of in-situ resident cell populations, such as somatic cells present in scar tissue formed after myocardial infarction (MI) or the conversion of lineage-related cells such as glia cells toward a distinct neuronal cell type. Recently published data suggest a chemically induced extra-embryonic endoderm (XEN)-like state, thereby omitting a pluripotent stage [29]. Based on these XEN-like cells, neuronal and hepatocytic inductions were demonstrated [29].

This chapter provides an overview of recent progress in diverse (re-) programming strategies, with particular focus on cardiovascular subtype differentiation. We summarize common cell fate conversion concepts for various somatic cell types and draw attention to their advantages and disadvantages in comparison with PSC strategies, considering the foremost priority of all presented projects to be the achievement of mature and physiologically functional cells that reflect, as closely as possible, the properties of their natural counterparts.

2 Direct Cell Fate Conversion of Somatic Cells

Direct reprogramming of patient-specific somatic cells offers enormous potential for individual screenings and patient-adjusted solutions for multifaceted dysfunctions. Moreover, feasible cell fate conversion of resident cell populations may reduce the risk of tumorigeneses and inflammation as well as avoid a potentially hazardous ex vivo cultivation step. Transplantation of the patient's own somatic cells or transdifferentiation of autologous resident cells can prevent immunological rejection, which enhances therapeutic outcome. Successful lineage conversions of all three embryonic germ layer derivatives have been achieved in recent decades (Fig. 1) and are discussed in brief in Sects. 2.1–2.3, with tabular overviews of recently published reports (Tables 1, 2, and 3). The generation of a desired cell type can be achieved through cell fate conversion within one germ layer or across germ layers (ectodermal, endodermal, and mesodermal), without passing an intermediate pluripotent stage, based on forced expression of lineage-specific TFs, noncoding RNAs, and signaling pathway modulators.

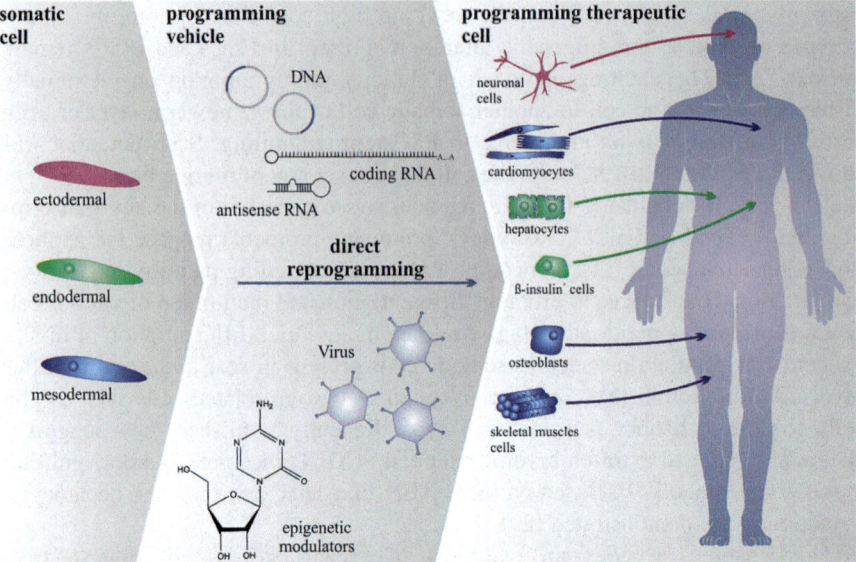

Fig. 1 Direct reprogramming strategies using somatic cell fate conversion within or across germ layers. Terminally differentiated cells from all three germ layers (*left*) are transformed into the desired therapeutic cell type (*right*) using various DNA, RNA, or nucleic acid-free reprogramming vehicles (*middle*)

2.1 Ectodermal Lineage

2.1.1 Neuronal Cells

Distinct neuronal populations are required in order to regenerate degenerated brain tissue and are defined through their receptor occurrence and release of neurotransmitters such as GABA, dopamine, or glutamate. Several approaches aim to exploit non-neurogenic astroglia [30, 33, 35, 39] (Table 1), which are a cell population of glial cells, the second major cell type after neurons in the brain. Other groups have explored direct reprogramming of somatic cell fates across germ layers using mesodermal fibroblasts as starting material [31, 32, 34, 36–38] (Table 1). Both strategies have yielded promising results, including inhibitory and excitatory neurons with important characteristics such as functional synapse formation and physiological activity. Thereby, although strategies based on small molecules yield relatively heterogeneous cell populations, they demonstrate the feasibility of chemically induced reprogramming without any genetic manipulation of human or mouse cells, even across germ layers [35–38]. However, selective expression of distinct neurogenic fate determinants, such as achaete-scute homolog 1 (Ascl1), basic helix-loop-helix (bHLH) TF family, neurogenic differentiation 1 (NeuroD1), and neurogenin-2 (Ngn2), seems to be more suitable for precise conversion toward a distinct neuronal subtype [30, 31, 33, 34, 39].

Table 1 The most prominent ectodermal lineage conversion strategies

Literature	Host	Original cell type	Target cell type	Modulator	Efficiency
Neuronal cells					
Heinreich et al. [30]	Mouse	Postnatal cortical astroglia	GABAergic neuron Glutamatergic neuron	Dlx2 or Ascl1, Dlx2 Ngn2	Functional synapses
Caiazzo et al. [31]	Mouse Human	MEF (mesoderm) Adult healthy and Parkinson's disease fibroblasts (mesoderm)	Dopaminergic neuron	ASCL1, LMX1A, NURR1	Dopamine release Spontaneous electrical activity
Yoo et al. [32]	Human	NFF (PCS-201-010); adult DF (mesoderm)	Excitatory and inhibitory neuron	miR-9/9*, miR-124, NEUROD2, ASCL1, MYT1L	Heterogeneous population with marker expression of excitatory (*VGLUT1*, *SLC1A2*) and inhibitory (*GAD67*, *DLX1*) neurons, functional activity
Guo et al. [33]	Mouse Mouse Human	NG2 Postnatal cortical astroglia, NG2	GABAergic neuron Glutamatergic neuron Glutamatergic neuron	NeuroD1	Neuronal marker expression (NeuN; DCX) functional activity with large GABA, glutamate, and NMDA currents
Colasante et al. [34]	Mouse Human	MEF (mesoderm) Lung fibroblasts (MRC-5) (mesoderm)	GABAergic neuron GABAergic neuron	Foxg1, Sox2, Ascl1, Dlx5, Lhx6 Foxg1, Sox2, Ascl1, Dlx5, Lhx6, Bcl2	Functional activity similar to cortical interneurons Functional synapses GABA release 25% GABA⁺ cells Expression of TUBB3 and PV
Zhang et al. [35]	Human	Astroglia (HA1800)	Neuron	LDN193189, SB431542, TTNPB, Tzv, CHIR99021, VPA, DAPT, SAG, Purmo	Survival >5 months in culture Functional synapses Synchronous burst activities
Li et al. [36]	Mouse	MEF (mesoderm)	Neuron	Forskolin, ISX9, CHIR99021, SB431542, I-BET151	>90% TUJ1⁺ cells after 16 days Functional synapses Action potential

(continued)

Table 1 (continued)

Literature	Host	Original cell type	Target cell type	Modulator	Efficiency
Hu et al. [37]	Human	Adult fore-skin fibro-blasts (FS090609) (mesoderm)	Neuron	VPA, CHIR99021, Repsox, forskolin, SP600125, GO6983, Y-27632	Neuronal marker expression (Dcx, Tuj1, Map2) Repetitive trains of action potentials after membrane depolarization
He et al. [38]	Mouse	MEF (mesoderm)	Neuron	DMEM/F12 supplemented with N2, bFGF, LIF, AA, and βMe	40% Tuj1$^+$ cells after 16 days voltage-gated potassium current, no sodium current, or spontaneous postsynaptic current
Rivetti et al. [39]	Mouse	Astroglia	Dopaminergic neuron	NEUROD1, ASCL1, LMX1A, miR218, AA, SB431542, LDN193189, CT99021	Efficiency 16%, appropriate mid-brain markers and excitability

2.2   Endodermal Lineage

2.2.1   Hepatocytes

Chronic, alcoholic, or fatty liver diseases engender liver fibrosis, making current cell therapy strategies ineffective because of impaired engraftment [83]. Hence, strategies for direct reprogramming of resident myofibroblasts (MFs) offer great potential for future therapeutic options. Surprisingly, a cell fate conversion of profibrogenic MFs toward hepatocyte (HC)-like cells across germ layers does not seem to be a major obstacle. However, all attempts harnessed the potential of forced exogenous overexpression of lineage-specific TFs, especially Hnf1a (which is always required) [40–46]. These strategies include two integration-free methodologies, transfection with synthetic modified mRNA [43] and an episomal delivery system [44]. Most HC-like cells display the functional characteristics of mature HCs such as albumin (ALB) secretion, cytochrome P450 activity, and storage of glycogen. Moreover, transplanted or in vivo generated HC-like cells are able to reduce liver fibrosis and restore liver function, which leads to extended survival [40–42, 44–46]. However, the generated cell populations are not fully mature HCs and demonstrate expression of immature cell markers such as α-fetoprotein (AFP) [43].

Table 2 The most prominent endoderm lineage conversion strategies

Literature	Host	Original cell type	Target cell type	Modulator	Efficiency
Hepatocytes					
Huang et al. [40]	Mouse	TTF (mesoderm)	Hepatocyte (HC)-like cells	Gata4, Hnf1α, Foxa3, knockdown of p19^{Arf}	Epithelial morphology repopulation of livers in mouse model, 23% Alb$^+$ cells
Huang et al. [41]	Human	FLF (mesoderm)	HC-like cells	HNF1A, HNF4A, FOXA3, SV40 large T antigen	Cytochrome P450 enzyme activity, biliary drug clearance, 20% ALB$^+$ cells
Du et al. [42]	Human	hEF (mesoderm)	HC-like cells	HNF1A, HNF4A, HNF6, CEBPA, ATF5, PROX1, p53-siRNA, C-MYC	90% ALB$^+$ cells, cytochrome P450 enzyme activity
Simeonov and Uppal [43]	Human	NNF (mesoderm)	HC-like cells	HNF1A, FOXA1, FOXA3, HNF4A	Expression of, e.g., AFP (immature HCs), ALB (mature HCs), APOA1, SERPINA1
Kim et al. [44]	Mouse	MEF (mesoderm)	HC-like cells	Gata4, Hnf1a, Foxa3	Expression of, e.g., Afp, Alb, Gata4, Hnf4a, E-cadherin ~70% glycogen storage ~50% xenobiotic metabolic activity Alb secretion
Song et al. [45]	Mouse	MF (mesoderm)	HC-like cells	FOXA3, GATA4, HNF1A, HNF4A	In vivo reduction of liver fibrosis, typical primary hepatocyte marker expression, storage of glycogen, uptake of LDL, secretion of Alb, cytochrome P450 (CYP1A2 and 3A) activity
Rezvani et al., [46]	Mouse	MF (mesoderm)	HC-like cells	Foxa1, Foxa2, Foxa3, Gata4, Hnf1a, or Hnf4a	Cytochrome P450 (CYP) gene expression, Alb secretion, CYP3A activity, urea production

(continued)

Table 2 (continued)

Literature	Host	Original cell type	Target cell type	Modulator	Efficiency
Insulin⁺ β-cells					
Zhou et al. [47] Cavelti-Weder et al. [48]	Mouse	Pancreatic exocrine cells	β-cell-like cells	Ngn3, Pdx1, MafA	Insulin secretion in vivo 92.8% Glut2$^+$ cells 85.3% Nkx2.2$^+$ cells 85.9% Nkx6.1$^+$ cells
Banga et al. [49]	Mouse	SOX9$^+$ HCs	β-cell-like cells	Ngn3, Pdx1, MafA	Dense core granule glucose-sensitive insulin secretion
Lemper et al. [50]	Human	Pancreatic exocrine cells	β-cell-like cells	Activated MAPK and STAT3	50–80% NGN3$^+$ cells Insulin secretion
Zhu et al. [51]	Human	NFF (CRL-2097); DF (mesoderm)	β-cell-like cells	OCT4, SOX2, KLF4, shRNA (p53), EGF, bFGF, CHIR99021, ActivinA, A83–01, nicotinamide, forskolin, dexamethasone, exendin, Compound-E, vitamin C, and BayK-8644	7% C-peptide$^+$ cells Expression of PDX1, NKX6.1, and NKX2.2 Glucose-stimulated insulin secretion in vivo
Yang et al. [52]	Mouse	HCs	β-cell-like cells	Pdx1, Ngn3, MafA	Glucose-stimulated insulin secretion in vivo

2.2.2 Insulin-Positive β-Cells

Diabetes mellitus is associated with the loss or dysfunction of insulin-secreting β-cells (a subpopulation of endocrine islet cells) in the pancreas. These cells have therefore aroused great interest for cell replacement therapies. One of the underlying concepts addresses a cell fate conversion of (endodermal) germ layer-derived pancreatic exocrine cells to pancreatic endocrine islet cells [47, 48, 50]. A patient's autologous cell source for the endodermal germ layer could be HCs from the liver [49, 52]. Both approaches are dependent on the forced exogenous overexpression of the lineage-specific TFs neurogenin 3 (Ngn3), insulin promoter factor 1 (Pdx1), and v-maf musculoaponeurotic fibrosarcoma oncogene family, protein A (MafA) [47, 49, 52]. Direct reprogramming across germ layers seems to be a possible alternative, using fibroblasts [51]. However, this approach requires introduction of the stem cell inducers Oct4, Sox2, and Klf4, with a stepwise protocol to generate initially endodermal progenitor cells, followed by posterior foregut-like specified

Table 3 The most prominent mesoderm lineage conversion strategies

Literature	Host	Original cell type	Target cell type	Modulator	Efficiency
Bone formation					
Yamamoto et al. [53]	Human	GF, DF	Osteoblast (OB)-like cells	RUNX2, OSTERIX, OCT4, L-MYC	~80% ALP activity Calcium deposition
Li et al. [54]	Human		OB-like cells OC-like cells	RUNX2, dexamethasone, CHIR99021, forskolin	Endogenous Runx2 and Osterix expression ALP activity OBs: mineralized nodule deposition; bone formation OCs: ramifications extended from the cell body
Skeletal muscle cells					
Davis et al. [28]	Mouse	MEF (C3HT10½)	Myocytes	MyoD, 5-azacytidine	Expression of, e.g., Mhc, Mlc2
Warren et al. [55] and Hausburg et al. [56]	Mouse	MEF (C3HT10½)	Myocytes	MyoD – modified mRNA	MyHC expression
Bichsel et al. [57]	Mouse	MEF (C3HT10½)	Myocytes	MyoD – protein (bacterial injection)	44% des⁺ cells 37.6% myog⁺ cells
Kim et al. [58]	Human	Urine-derived cells	Myocytes	MyoD – lentiviral transduction	Upregulation of *DES, MYOG, MYH3, ACTN2, DMD*
Horio et al. [59]	Human	Skin fibroblasts	Myocytes	MyoD – adenoviral transduction	Cell fusion and high motility Ca²⁺ release Expression of, e.g., myog, dystrophin

(continued)

Table 3 (continued)

Cardiomyocytes

Literature	Host	Original cell type	Target cell type	Modulator	Efficiency
Ieda et al. [60], Chen et al. [61], Qian et al. [62], Inagawa et al. [63], Qian et al. [64], Wang et al. [65]	Mouse	TTF, CF	Cardiomyocyte (CM)-like cells	Gata4, Mef2c, Tbx5 (GMT)	Stoichiometry of G, M, T protein expression influences reprogramming efficiency 30% or 35% $cTnT^+$ cells 3% or 20% αMHC^+ cells 10–15% efficiency
Song et al. [66]	Mouse	CF, TTF	CM-like cells	Gata4, Mef2c, Tbx5, Hand2 (GMTH)	GMTH: 6.8% $cTnT^+/\alpha\text{-}MHC^+$ GMT: 1.4% $cTnT^+/\alpha\text{-}MHC^+$
Jayawardena et al. [67, 68, 69]	Mouse	NNF, TTF	CM-like cells	miR-1, miR-133, miR-208, miR-499, JAK inhibitor I	28% αMHC^+ cells enhanced cardiac function in mouse model
Nam et al. [70]	Human	NFF and adult fibroblasts	CM-like cells	GATA4, HAND2, TBX5, MYOCARDIN miR-1/−133 Culture time: 4–11 wk	~35% tropomyosin$^+$ cells ~20% $cTnT^+$ cells
Hirai et al. [71, 72]	Mouse	TTF, MEF (B6; 129S4)	CM-like cells	M_3 domain of mouse MyoD fused on carboxy-terminus of Mef2c, Gata4, Hand2, Tbx5 GSK126 (day 1–4), UNC0638 (day 3–7)	Reprogramming efficiency: MM_3-GHT 3.5% (>15-fold increase) MM_3-GHT + GSK126: Further 2.1-fold increase compared with control (most efficient combination) MM_3-GHT + UNC0638: Further 2-fold increase compared with control
Wang et al. [73]	Mouse	MEF, TTF	CM-like cells	Oct4, SB431542, CHIR99021, parnate, forskolin	Expression of Myh6, Tnnt2, Ryr2, Gata4, Nkx2–5, cTnT, Cx43 Ventricular-like action potential

Fu et al. [74]	Mouse	MEF	CM-like cells	Two-stage protocol: day 0–16: CRM AA, CHIR99021, RepSox, Forskolin, VPA, Parnate, TTNPB Day 17-end: CHIR99021 PD0325901, LIF, insulin	Morphology: spindle shape, rod shape, or round shape Spontaneously beating activity that increases from day 8 Cardiac marker expression of Mef2c, α-actinin, Gata4, cTnT, Nkx2.5, α-MHC, N-cadherin, Cx43, cTnI Action potential of atrial- and ventricular-like CMs
Zhao et al. [75]	Mouse	MEF	CM-like cells	Gata4, Hand2, Mef2c, Tbx5, miR-1/−133, Y-27632, Thiazovivin, SR-3677, A83-01	~60% cTnT$^+$ cells ~60% α-actinin$^+$ cells
Zhou et al. [76]	Mouse	MEF, CF, TTF	CM-like cells	Gata4, Hand2, Mef2c, Tbx5, Akt1	Spontaneously beating activity: MEFs > day 7 (50% > day 21), CFs > day 14, TTFs > day 21 Responsive to β-adrenoreceptor pharmacologic modulation, polynucleated, and hypertrophic
Yamakawa et al. [77]	Mouse	MEF, TTF	CM-like cells	Gata4, Mef2c, Tbx5, Hand2, FGF2, FGF10, VEGF (FFV)	FFV at late differentiation phase promotes reprogramming 9% beating cells
Talkhabi et al. [78]	Mouse	MEF	CM-like cells	Oct4, Sox2, Klf4, cMyc, AA	~40% GATA4$^+$ cells ~12% αMHC$^+$ cells
Park et al. [79]	Mouse	MEF, TTF	CM-like cells	Forskolin, A-8301, SC1, CHIR99021, (±)-BayK-8644, FGF2, AA, ITS	~27% cTNT$^+$ cells ~3% cTNT$^+$ cells
Cao et al. [80]	Human	HFF	CM-like cells	CHIR99021, A83-01, BIX01294, AS8351, SC1, Y27632, OAC2, SU16F, JNJ10198409	~7% cTnT$^+$ cells Expression of Cx43, HCN4, cTNI, ANF, MLC2v Ventricular-like action potentials

(continued)

Table 3 (continued)

Literature	Host	Original cell type	Target cell type	Modulator	Efficiency
Zhou et al. [81]	Mouse	NNCF	CM-like cells	Gata4, Mef2c, Tbx5 shRNA of 35 selected components of chromatin modifying or remodeling complexes	Bmi1 downregulation significantly enhanced CM generation
Mohamed et al. [82]	Mouse	NNCF	CM-like cells	Gata4, Mef2c, Tbx5 SB431542 XAV939	Eightfold increased reprogramming efficiency Beating cells 1 week after reprogramming enhanced cardiac function in mouse model

progenitor cells, and, finally, pancreatic β-like cells via temporal application of several chemical compounds [51]. All attempts can serve as a good basis for innovative cell therapies based on the observed islet β-cell-specific gene expression profile and efficient glucose-stimulated insulin secretion in vivo.

2.3   Mesodermal Lineage

2.3.1   Bone Formation

Mineralization is a prerequisite for the regeneration of weakened bone and therefore osteocytes (OCs) are the most common cell type in mature bones. OCs are derivatives of osteoblasts (OBs), which can be efficiently generated via direct reprogramming of human fibroblasts with the combination of the TFs Oct4, L-Myc, and the osteoblast-specific TFs Runt-related transcription factor 2 (Runx2) and Osterix using in vitro osteogenic culture medium [53]. These highly mature and osteocalcin-producing cells exhibit a gene expression profile similar to that of normal human OBs. Li et al. reduced the number of necessary TFs and induced a cell fate conversion toward OCs und OBs using a chemical cocktail by activating Wnt and cAMP/PKA pathways in combination with Runx2 [54]. Both cell types demonstrated cell type-specific characteristics. It is particularly interesting to know whether sophisticated timing of chemical cocktail exposure could be used to achieve the desired maturation grade, thereby yielding either pure OBs or OCs.

2.3.2   Skeletal Muscle Cells

Many hereditary and refractory diseases could benefit from clinically relevant myocytic sources. MyoD was discovered to be a sufficient TF for a direct cell fate switch from fibroblasts to myoblasts [28] by activating a feed-forward circuit to regulate muscle gene expression. Several groups have since explored optimal protocols for efficient MyoD introduction into cells, such as genome integration using adenoviral [59] and lentiviral [58] transduction or cytoplasmic application using either modified mRNA [55, 56] or MyoD protein via bacterial protein injection system (the so-called type III secretion system) [57]. Notably, a literature search led to no hits for using a chemical cocktail as an alternative to MyoD application.

2.3.3   Cardiomyocytes

The foremost reasons for MI are hypertension and narrowing of the coronary arteries caused by arteriosclerosis [84]. This results in occlusion of coronary vessels and subsequent undersupply of the affected tissue [85], which culminates in CM death [86]. The process underlying cell death varies and is not exactly defined [86], whereby death by apoptosis [87–89], necrosis [87, 90–92], and in association with autophagy [93–95] have been described. Thus, 25% of the human left ventricle (0.5–1 billion CMs) can be destroyed by MI within a few hours [96, 97].

In response to the injury, a cascade of numerous biochemical and mechanical processes are stimulated, which, in consequence, provoke cardiac dysfunction and loss of functionality [98–108]. Wall thinning, collagen degradation, and ventricular dilatation are incipient steps [104]. Circulating cells migrate into the wound area and, in combination with an accumulation of extracellular matrix proteins in the cardiac interstitium, lead to cardiac fibrosis [109, 110]. The emerging scar tissue provokes reduced systolic function through wall stiffening [100]; moreover, faulty electrical coupling is particularly noticeable due to the massive cell loss [98, 100].

In addition, the regeneration potential of CMs in the human heart is extremely low, with a turnover rate of 1% at the age of 25, which decreases to 0.45% by the age of 75 [1].The percentage of CMs situated in mitosis and cytokinesis is highest in infants, suggesting significant regenerative potential of the myocardium in children and adolescents [2]. Thus, the negligible regeneration capacity of the myocardium of the mainly affected age group results in unfeasible functional repair and deleterious remodeling of damaged tissue after MI. So far, resident cell populations such as cardiac progenitor cells (CPCs) or preexisting CMs are the only, but unfortunately insufficient, sources for myocardial regeneration after injury [2, 111, 112].

In this context, cell fate conversion of resident cardiac fibroblasts (CFs) enables the most efficient direct reprogramming toward CMs, probably within the same germ layer. Therefore, current research widely refers to easily available murine neonatal CFs because of their heterogeneity, plasticity, and resistance to the hypoxic environment of the injured myocardium [103, 113, 114] (Table 3). Another prominent source is murine embryonic fibroblasts (MEFs) [71, 72, 74–77]; however, neither of these two cell types is easily accessible from humans. Therefore, until now, only human foreskin fibroblasts (HFFs) have been used [70, 80]. Notwithstanding, it is possible that false-positive results on CFs and MEFs can arise from contamination with CMs. Therefore, specific mouse models have to be established marking descendants of nonmyocytes to ensure isolation of pure fibroblast populations, as shown for example by Qian et al. [62].

A cell fate switch of fibroblasts toward CM-like cells is mostly achieved through forced exogenous overexpression of lineage-specific TFs, whereby Gata4, Tbx5, and Mef2c is the most frequently used TF combination [60–65] as part of a more complex composition [71, 72, 75–77, 81, 82]. However, the obtained marker gene expression patterns vary widely among laboratories and starting materials; for example, 30% [60] or 35% [61] cTnT$^+$ cells, 10–15% α-actinin$^+$ [62], 3% [63] or 20% [64] αMHC$^+$ cells have been reported. Furthermore, the cells display only marginal similarity to mature CMs at the molecular and electrophysiological levels [61]. This weak efficiency could be caused by the use of suboptimally designed constructs; enhanced programming efficiency was demonstrated with a tailored ratio of protein expression (i.e., higher protein levels of Mef2c together with lower levels of Gata4 and Tbx5) [65]. Several approaches aim to enhance reprogramming efficiency and maturation of induced CMs (iCMs) by adding further TFs of the cardiac lineage such as Hand2 [70–72, 75, 76] as well as signaling modulators such as inhibitors of TGF-β (A83–01, SB431542) [75, 82], Wnt (XAV939) [82], and ROCK (SR-3677, Thiazovivin, Y-27632) [75]. A further improvement could be achieved in vivo compared with in vitro, as demonstrated for murine cardiomyogenic differentiation by Qian et al. [62] and Mohamed et al. [82], which underlines the need for a cardiogenic microenvironment for efficient and distinct cell fate conversion toward mature CMs. In vivo-generated iCMs exhibit a more mature sarcomeric phenotype and a reaction to electrical stimulation similar to that of adult ventricular CMs [62]. Moreover, in vivo strategies result in enhanced cardiac function [62, 82] with muscle restoration in the infarct region, as displayed by magnetic resonance imaging (MRI).

In addition, several procardiogenic microRNAs have been identified [115–117], which are deployed in combination with TFs [70, 75] or as sole modulators [67–69]. MicroRNAs such as microRNA-1 interact with myogenic TFs such as serum response factor, Mef2c, MyoD, or Nkx2.5 as repressors and cooperators in a regulatory loop [118–120]. MicroRNA-1 has a negative impact on the Notch signaling pathway through direct repression of Dll1 [121] and its downstream effector Hes1 [122], which results in expression of Gata4, Nkx2.5, and Myog. A combination of miR-1/-122/-208/499 and JAK inhibitor I achieved a cell fate switch toward a cardiogenic phenotype with 28% αMHC$^+$ cells in vitro and improved cardiac function in vivo [67–69].

Furthermore, Fu et al. demonstrated a successful cell fate switch for the first time by using a defined chemical cocktail to convert MEFs to iCMs with spindle shape, rod shape, or round morphologies, thereby avoiding genome integrative modulation [74]. The generated cells exhibited a heterogeneous action potential profile of atrial- and ventricular-like cells. In the same year, Park et al. [72] reported another chemical cocktail to induce cardiogenic conversion using five enhancers for iPSC induction, achieving ~27% cTNT$^+$ cells (MEFs) or 0.84–2.82% cTNT$^+$ cells (tail tip fibroblasts; TTFs). In this regard, effective time windows for every chemical modulator (e.g., signaling pathway activators/inhibitors, epigenetic regulators) need to be defined to achieve the best results [72].

To date, the obtained iCMs display a highly immature and inhomogeneous population lacking the terminal structural and electrophysiological characteristics of adult CMs. Moreover, despite the many advantages of directly reprogramming resident CFs, the ultimate consequences of massive fibroblast (or other cardiac cell type)-to-myocyte conversion remain unknown and may possibly be detrimental to heart function [123, 124].

3 Multipotent and Pluripotent Stem Cell-Based Differentiation Strategies for the Cardiovascular Lineage

Because the direct reprogramming concepts described are still incapable of yielding pure matured cardiac subtypes, stem cell-based strategies remain of crucial importance.

3.1 Mammalian Multipotent Stem Cell-Based Approaches

As mentioned, sole application of ASCs did not lead to a significant improvement of LVEF after MI [13, 14]. However, modification of mesenchymal stem cells (MSCs) by enhancement of cell survival and proliferation as well as stimulation of paracrine factor secretion and neoangiogenesis, thereby promoting cardiac repair, may provide a possible solution [125].

Certainly, directed differentiation of multipotent stem cells could offer greater benefits. The required stem cells for such strategies can be obtained from various sources, such as bone marrow (BM-MSCs) [122, 126–132], adipose-tissue (ADSCs) [133–138], or dental follicles (DFSCs) [139]. Also, diverse progenitor cells can be used, as demonstrated for endothelial progenitor cells (EPCs) isolated from peripheral blood of patients with acute MI and from umbilical cord blood [140] or cardiac progenitor cells (CPCs) isolated from fetal hearts, with either c-kit$^+$ [141, 142] or Sca-1$^+$ [142, 143] cell populations.

Cardiogenic-directed differentiation could be induced using various exogenous manipulation strategies. Most of the published data are based on modulations using methylation inhibitors (e.g., 5-azacytidine) and histone deacetylase inhibitors (e.g., trichostatin A) [126, 128, 129, 131–134, 137, 140], which underlines the high impact of epigenetic alteration as an important target for cell fate conversion. Another promising and much-discussed concept is the use of in vitro co-culture with isolated neonatal CMs [127, 130, 137]. However, more recent reports suggest that cardiac marker expression of MSC and CD34$^+$ progenitor cell derivatives occur on the basis of cell fusion with recipient CMs rather than by differentiation in vivo [144, 145]. To avoid inconclusive results, indirect co-culture using inserts was tested; nonetheless, direct cell–cell contact yielded better results and even led to ADSC-derived spontaneously beating CM-like cells [137]. Further strategies are based on forced exogenous overexpression of either cardiogenic-specific TFs, such as Shox2 [130], Gata4 [127], and Nkx2.5 [127] or a TF cocktail [134]. Other strategies use noncoding RNAs, including microRNA [122] and long noncoding (lnc)RNAs [132], or stimulation with media supplements such as growth factors [129, 135, 136] or chemicals such as ascorbic acid [136] and suberoylanilide hydroxamic acid [139]. The outcomes vary, although expression of specific cardiac markers such as desmin, cardiac actin, and troponin has been demonstrated. However, none of these approaches is currently sufficient to generate a pure cardiomyocytic population.

3.2 Human Pluripotent Stem Cell-Based Approaches

PSCs are a highly valuable cell source for studying key cellular and molecular programs of early embryonic development, including that of the heart. This topic is addressed in detail by Kempf and Zweigerdt in another chapter of this volume

[146]. In general, because of lack of a perfect imitation of the endogenous micro-environment, including topographical, electrical, adhesive, mechanical, biochemical, and cell–cell interaction cues [147], entirely mature and fully physiological functional CM differentiation is still an unmet goal in vitro, even when using ESCs or iPSCs [148]. Nevertheless, PSC differentiation concepts are mainly inspired by natural processes during embryonic development, thereby directing cell fate alongside time-, space-, and signaling-dependent patterns to overcome obstacles accompanying species specification and interpersonal variations. To prevent unnecessary costs and delays in the future, it is of interest to the scientific community and the public to introduce a highly standardized and uniform analysis system. Further success could be monitored and information better compared. At present, multitude TF or surface marker expressions are addressed at diverse time points.

Recent studies of human PSCs present chemical-based rather than DNA-integration-based strategies, which will probably facilitate translation from bench to bedside. Several compound cocktails intervene by activation or inhibition of lineage-relevant signaling pathways, including ROCK inhibition through Y-27632 [149–152] and H1152 [153]; activation of Wnt signaling through CHIR99021 [149, 150, 152, 154–156] and Wnt inhibition through IWR1 [149, 152, 154], IWP2 [154] and IWP4 [150]; Activin/Nodal/TGF-β activation through LY294002 [154]; and application of BMP4/ActivinA [151, 153, 154, 157, 158] and Activin/Nodal/TGF-β inhibition through SB-431542 [151].

However, with nucleic acid-free concepts it is more difficult (but not impossible) to purify the cell population of interest. A selection of human CM-like cells has been achieved via mitochondria-specific fluorescent dyes (99% α-actinin$^+$ cells [159]), antibodies against CM-specific markers such as signal-reduced protein alpha (SIRPA; 98% cTnT$^+$ cells [160]), elastin microfibril interface 2 (EMILIN2; no qualitative statement about α-actinin$^+$ or cTnT$^+$ cell yield [161]), vascular cell adhesion molecule 1 (VCAM1; 95% cTnT$^+$ cells [162]), or utilization of the sugar/lactate metabolism (98% α-actinin$^+$ cells [163], 90% cTnT$^+$ cells [153], 83.3% cTnT$^+$ cells [152]).

Nonetheless, PSC-derived CMs are often a mixture of nodal-, atrial-, and ventricular-like phenotypes, as revealed by electrophysiological and pharmacological studies [164]. Therefore, several approaches now proceed from classical cell cultivation toward co-culture concepts such as AKT-activated endothelial cells, which led to an improvement of Nkx2.5$^+$ cells as well as faster beating activities compared with hESCs cultured on Matrigel [165], matrix–cell composites such as bioartificial cardiac tissue (BCT; cells plus liquid collagen type I plus Matrigel) [166], or cardiac extracellular matrix [167]. To supply endogenous tissue with blood capillary networks, Akashi's group developed a vascularized 3D iPSC-CM tissue, which provided comprehensive data for drug screening [168]. Another approach uses co-culture of human cardiac microvascular endothelial cells (hCMVECs) and hMSCs in combination with human induced pluripotent stem cell-derived embryonic cardiac myocytes (hiPSC-ECMs) to generate vascularized cardiac tissue [169]. A report by Eder et al. confirmed the considerable importance of human 3D heart tissue obtained using iPSC technology to overcome species-dependent discrepancies in CM behavior [170].

However, to ensure secure cell replacement therapies, drug development, and disease modeling using PSC-derived CMs, much remains to be done, such as assuring compliance with good manufacturing practice (GMP) standards and reliable integration-free protocols with the possibility of large-scale production [171].

4 Programming of Cardiac Conduction System Cells

In addition to cardiac diseases such as MI and ischemia, cardiac arrhythmias can also impair life, especially if the symptoms are pronounced and need treatment. Arrhythmias can usually be traced back to malfunction of whole parts or of some cells of the cardiac conduction system (Fig. 2), which typically induces the "sick sinus syndrome" (SSS). This term describes a collection of diverse signs and symptoms that specify a disease condition and can result from various causes [172]. In SSS, the sinus node no longer generates normal cardiac impulses or there is no proper conduction throughout the heart. The resulting arrhythmias can include sinus bradycardia, sinus pauses, sinus arrest, and sinoatrial exit blocks. In ~50% of the cases, alternating bradycardia and tachycardia occur [172–175]. Various other symptoms such as lightheadedness, syncope, fatigue, and palpitation can also be observed [172]. SSS occurs predominantly in the elderly, but is prevalent at

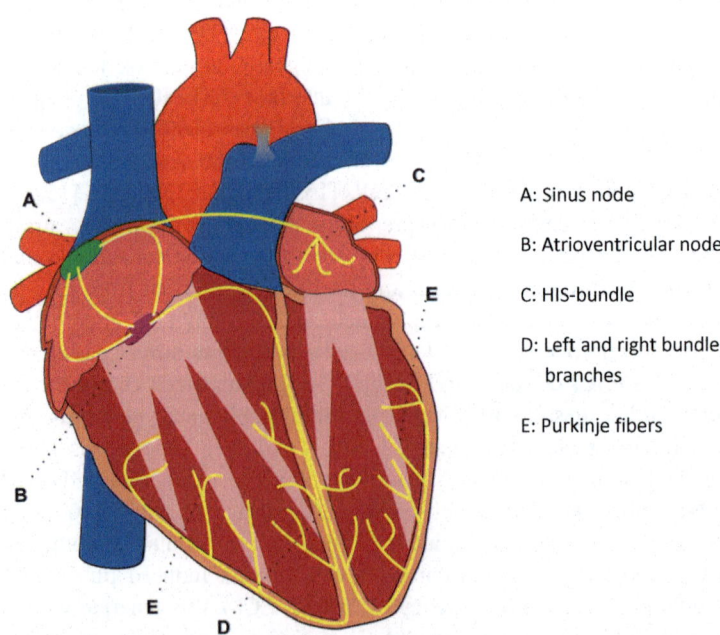

A: Sinus node

B: Atrioventricular node

C: HIS-bundle

D: Left and right bundle
    branches

E: Purkinje fibers

Fig. 2 Major components of the cardiac conduction system in mammals, shown in *green*, *violet*, and *yellow* against the background of a schematic longitudinal section through the heart

all ages [172–175]. Whereas SSS of young adults and children is commonly a consequence of post-operative atrial trauma or a genetic issue, more causes exist in the elderly, such as reduced cell number in the sinus node [174], reorganization of the sinus node caused by several heart defects [172], or coronary artery disease. The sinus node is highly energy consuming and, consequently, even temporary interruption of its blood supply via the right coronary artery (provoked by ischemia or MI) causes permanent malfunction [172, 175]. Familial SSS, caused by genetic mutations, often originates from aberrations in one of three genes [172]. Two of these, the hyperpolarization-activated cyclic nucleotide channel 4 (HCN4) and the sodium channel, voltage-gated, type V, alpha subunit (SCN5A) are essential for formation of transmembrane ion exchange and therefore highly relevant for generation of proper action potentials. The third gene, myosin, heavy chain 6, cardiac muscle, alpha (MYH6) has a crucial role in formation of the contractile apparatus [172].

Of note, there is currently only one long-term supportive therapy option available, namely the implantation of an electrical pacemaker. This device relieves symptoms, offers a better quality of life, and improves survival in certain cases [176]. SSS is one of the major indications for pacemaker implantation (30–50% of all cases) [172–175, 177]. However, this therapy also has some disadvantages, such as risk of infection, limited battery life, tearing of leads, and electromagnetic interference [176]. A possible solution to these shortcomings is a biological pacemaker, and a number of promising approaches in this direction have appeared in recent years [177]. All these attempts aim at altering the characteristics of diverse target cell types into a more pacemaker-like phenotype, either by converting resident cells of the heart [178–180] or by preprocessing cells in vitro and then transplanting them into the heart [151, 181–183]. An additional important benefit of the availability of highly pure in vitro-generated pacemaker cells would be their use for personalized in vitro drug testing.

4.1 Direct Reprogramming of Somatic Cells

Early approaches toward a biological pacemaker attempted to reprogram cells via exogenous introduction of ion channels into target cells to alter repolarization. The basis for this concept was pioneered by Johns et al. [184], who expressed K^+ channels in working myocardial cells, leading to shortened action potential duration (APD). Based on this work, Marban's group used dominant-negative mutants of the Kir2 gene family [185–187], thereby inhibiting the inward rectifier potassium current (I_{K1}) to create a de novo cardiac pacemaker, rendering ventricular myocardium spontaneously active [188]. When the construct was adenovirally administered to guinea pigs, it partially sufficed to elicit pacemaker activity in ventricular myocytes. However, it was subsequently found that Kir2.1 mutant overexpression not only destabilized the resting membrane potential, but also caused prolongation of APD [189]. This could potentiate early afterdepolarizations

(EADs), with the risk of ventricular arrhythmias [190, 191]. Therefore, additional coexpression of HERG was proposed in order to keep the APD short, but not interfere with the resting membrane potential destabilizing effect of the mutated Kir2.1 variant [192].

As described above, HCN channels are a crucial mediator of cardiac pacemaking [193]. Therefore, instead of inhibiting K$^+$ channels, overexpression of HCN channels was also addressed, with the goal of eliciting ectopic pacemaker activity in small and large animal models [194–196]. Moreover, as an alternative to wild-type HCN channels, synthetic pacemaker channels bearing a canonical voltage-dependent K$^+$ channel backbone were created with the goal of activating the channel on hyperpolarization, yet with nonspecific cation selectivity [197]. However, the engineered protein bears potentially immunogenic epitopes, possibly limiting long-term expression and translational potential.

In a more sophisticated approach, a fusion between host myocardial cells and syngeneic fibroblasts overexpressing HCN1 was created. The resulting hetero-karyons of myocytes and HCN1 fibroblasts revealed spontaneously oscillating action potentials [198]. In guinea pigs, electrocardiography showed biological pacemaker activity from 1 day after cell injection, which was stable for 2 weeks.

However, although proof of concept may have been obtained, the biological pacemaker concepts described above function only partially, most probably because they cannot truly replicate the complex physiology and morphological characteristics of genuine nodal cells [180]. More recent approaches rely on the use of cell fate determining transcription factors identified in the field of developmental biology to achieve "true" (re-)programming of target cells.

The programming factors need to be carefully selected for this highly specific purpose. TFs such as T-box 3 (Tbx3), T-box 18 (Tbx18), short stature homeobox 2 (Shox2), and ISL LIM homeobox 1 (Isl1) play a crucial role in development of PCs, although they are absent or strongly downregulated in other CM subtypes [199–203]. To date, only two TFs have been used for published studies, namely Tbx3 and Tbx18 [178–180, 201, 204].

As a transcriptional repressor during embryonic development, Tbx3 inhibits formation of the working myocardium by preventing expression of the responsible TFs. Thereby, Tbx3 imposes the pacemaker gene program, but does not seem to have a direct influence on the ultimate tissue architecture of the sinus node, which is organized by Tbx18 [203]. Two independent studies have investigated Tbx3 overexpression in mouse hearts. Yet, neither its expression in the atrial myocardium [201] nor tamoxifen-induced expression of Tbx3 in the whole working myocardium [178] led to fully functional PCs. Although both studies have promising aspects (i.e., SAN-specific markers are upregulated and atrial or ventricular markers downregulated), the resulting cells strongly differ from native PMCs. In particular, the HCN channels, which are essential for pacemaker function, appear to be misexpressed in the partially reprogrammed tissue [178, 201]. Therefore, Tbx3 alone does not seem sufficient for converting working myocardium into PCs.

Another study tested five different TFs with respect to their effect on neonatal rat ventricular myocytes (NRVMs; Shox2, Tbx3, Tbx5, Tbx18, and Tbx20)

[180]. Only Tbx18 significantly increased the number of spontaneously beating cultures. Although the resulting cells exhibited some pacemaker-like properties, such as altered morphology, enhanced HCN4 expression, and a pacemaker-like cellular automaticity, some other typical characteristics were lacking. In particular, the beating frequencies were still much less than a rat heart beat (95 bpm versus 350 bpm) even though they were twice that of control cells [180]. After these initial in vitro experiments, the effect of Tbx18 was also examined in vivo. The first experiments were performed in guinea pig hearts [180]. In a consecutive study, the Tbx18-expressing adenovirus was injected into the interventricular septum of pigs with a complete heart block [179]. Again, some pacemaker properties were observed after transduction, such as upregulated Hcn4 expression. More importantly, ectopic ventricular beats in guinea pig and pig hearts were induced [179, 180]. Likewise, the heartbeat of the pigs was autonomous and independent of an implanted electronic pacemaker [179]. A severe limitation of these studies is the transient effect of Tbx18 expression; cells isolated from guinea pigs after 6 weeks had lost their pacemaker-like morphology [180]. Similarly, the rapid recovery after electronic burst ventricular pacing in the Tbx18-transduced pigs had vanished after 2 weeks [179]. Although this study describes the first partially successful in situ induction of a biological pacemaker, the long-term effects of Tbx18 reprogramming of working myocardium remains to be determined [179, 180].

In a recent study, Tbx18 was expressed in the ventricular myocardium during fetal development using two independent Cre/loxP-mediated transgenic mouse models to investigate the potential of this factor to convert working myocardium into pacemaker cells [204]. Interestingly, right ventricular hypoplasia, atrial dilatation, and ventricular septal defects were found in the working myocardium, but no upregulation of the expression of SAN-related genes. Moreover, atrial and ventricular marker genes were also ectopically expressed, and downregulation of only a few chamber-specific genes was noticeable [204]. Correspondingly, no induction of a biological pacemaker by the expression of Tbx18 in the working myocardium was found in this study [204]. The contrasting outcomes of the two studies needs explanation The different utilization of Tbx18 in different species [179, 204] or the different expression time points in the heart (fetal [204] versus adult [179]) may partially underlie this phenomenon. These questions need to be answered to avoid any undesirable side effects before Tbx18 overexpression can be used for creation of a biological pacemaker in patients.

Beyond the direct reprogramming strategies of working myocardium in vivo, the promising studies of direct fibroblast conversion into spontaneously beating cells can serve as the basis for further development of the approach toward the generation of distinct CM subtypes such as pacemaker cells [205]. In this regard, experiments using the traditional CM reprogramming factors Gata4, Hand2, Mef2C, and Tbx5 revealed three cardiac cell types in the generated beating cells (atrial-like, pacemaker-like, and ventricular-like) [205]. Subsequently, transduction of 20 initial candidates and their successive omission, based on evaluation of their influence on Hcn4 expression, resulted in a final cocktail of four factors: Tbx5, Tbx3, Gata6, and either retinoic acid receptor gamma (Rarg), or retinoid X receptor alpha (Rxra). However, significant Hcn4 expression alone does not seem sufficient because no spontaneously beating cells were observed, nor were the cells excitable by a depolarization stimulus [205]. Although the described data are promising because they show that reprogramming fibroblasts into different cardiac subtypes is feasible, achieving such specified cells at high yield and purity is still far from being reliably established.

4.2   Conversion of Multipotent Stem Cells

Several studies have used various modifications of adult stem cells (ASCs). Mesenchymal stem cells [130, 206–217] derived from dogs [130, 207–210], rats [211, 218], rabbits [215–217], or humans [206, 212, 213] were primarily used, but there are also some data available from experiments with ADSCs [214, 219]. Most groups working with ASCs chose overexpression of a Hcn family member (see Sect. 4.1) [206, 208–213, 215–217] to drive the cell fate into the nodal phenotype, whereas others used TFs such as Shox2 [130, 207]. Combination with treatments such as 5-Azacytidin [214] or electric-pulse current stimulation (EPCS) [207, 208] is also used. Although the experimental setups vary, the scientific outcomes are comparable. Dependent on the experiment, the cells show some typical nodal cell properties. For example, the "funny" current (I_f) was measurable and could be enhanced with EPCS or isoproterenol and blocked with cesium [207–209, 213, 216]. Moreover, expression levels of characteristic pacemaker genes encoding proteins such as Cx45, Hcn4, Tbx3 increased, whereas expression of genes associated with the working myocardium, such as Cx43 and Nkx2–Nkx5, diminished [130, 207, 208]. A change in cell morphology to a more pacemaker-like phenotype was also observable [130, 207, 214]. Co-culture of the modified ASCs with neonatal myocytes, no matter the origin, led to higher beating frequencies of the neonatal myocytes in comparison to co-cultures with unmodified ASCs [130, 213, 216].

For in vivo testing, the cells were preferably transplanted into canine hearts [8, 12]. Two publications describe induction of a ventricular escape rhythm observed in ECG recordings after heart block induction [209, 213]. The first study reported a requirement for vagal stimulation [209]. Without stimulation, the only difference from transplanted control cells was the heart's beating frequency [213]. In all studies, the lack of autonomous activity of the cells was strikingly consistent [130, 206–218].

Another study describes the spontaneous activity of transformed stem cells derived from brown fat. Interestingly, this phenomenon seemed to reflect a spontaneous reaction to the cultivation media because no genetic modification had been applied [219]. Analysis of ultrastructure, proteome expression, electrophysiology, and pharmacology of the resulting beating cells revealed some pacemaker characteristics, but needed further examination, especially over longer culture time periods [219]. In summary, significant effort is required to achieve reprogramming of nodal cells from ASCs.

4.3 Programming and Differentiation of Pluripotent Stem Cells

Since mouse embryonic stem cells were differentiated into CMs for the first time in 1991 [220], the composition of different beating cells has been examined and attempts made at specific direction. In addition to morphology and expression patterns, electrophysiological properties have become increasingly important. Using traditional random differentiation protocols, the obtained cells represent all kinds of cardiac cells: nodal, atrial, ventricular, and immature CMs [221]. The spontaneous differentiation rate of nodal cells out of murine PSCs does not typically exceed 1%. Accordingly, it is of great interest to influence cell fate during differentiation. Moreover, although there have been great improvements regarding the differentiation of human PSCs into cardiac phenotypes via specific culture conditions [154, 222–225], the typical proportion of nodal cells under these conditions still needs to be defined.

There are currently three main strategies for enhancing the proportion of nodal cell type: stimulation via intrinsic culture conditions, enrichment via selection, or forced overexpression of specific TFs.

With respect to the first strategy, the small molecule compound EBIO has been postulated to enhance the formation of nodal cells from murine ESCs. Although the study reported some induction of the sinoatrial gene program and a reduction in the chamber specific gene program, the cells lacked a number of important properties. The beating frequencies were low, cells were not tested for their ability to pace chamber myocardium and mature pacemaker cells were not discriminated from the likewise spontaneously contracting early/intermediate cell type [226]. A more recent study, investigating the influence of EBIO on human PSC differentiation, revealed a better understanding of the underlying effect. Although supplementation with EBIO resulted in a dose-dependent enrichment of CMs with increased nodal- and atrial-like phenotypes, the effect was mainly attributed to reduced cell survival and thereby favored cardiac progenitor preservation [227].

Recently, a promising study described the generation of hPSC-derived pacemaker cells via a specific differentiation protocol combined with surface marker selection via SIRPA [151]. SIRPA is a cell-surface marker suitable for isolating populations of CMs from hPSCs [160]. The obtained cells fulfill a number of typical pacemaker characteristics; in particular, they are able to pace the host tissue after transplantation into rat hearts. Some points still need to be clarified; for example, the early/intermediated cell type was not taken into account despite the fact that the funny channel densities resembled those of immature cells and the action potential curves revealed clear plateau phases. Likewise, the characteristic Ca^{2+} release from the sarcoplasmic reticulum was not investigated and the specific morphology of single cells ("spindle" or "spider" cells) was not examined. Additionally, atypical expression of Cx43 and Cx40 was obvious from the data [151].

Another study based on surface marker selection used the activated leukocyte cell adhesion molecule (Alcam) during mouse ESC differentiation. Although the resulting cells showed pacemaker characteristics, selection via Alcam seemed extremely sensitive to time point and species because cell sorting at different time points resulted in different α-actinin expression [228]. Accordingly, an earlier study of human ESCs revealed an embryonic CM phenotype [229]. In addition, only about 10% of the selected cells retained Hcn4 expression after 3 weeks in culture. This could reflect maturation of initially Hcn4-positive early/intermediate CMs toward mainly working myocardial cells [228].

Two papers have addressed the potential of the transcription factor Shox2 to demarcate program nodal cells from murine ESCs. One group transfected a plasmid containing a neomycin resistance gene under control of the Shox2 promoter. Addition of neomycin during differentiation led to an almost pure population of Shox2-expressing cells. The analyzed marker patterns revealed some nodal characteristics and general cardiac markers. Although the cells were spontaneously active, their beating frequencies did not exceed ~120 bpm and no additional functional data were shown [230]. In the second study, Ionta et al. used Shox2 overexpression in mouse ESCs to force nodal cell lineage differentiation [181]. These cells were also spontaneously active, but the beating frequencies were below 80 bpm and therefore did not differ from those of wild-type-ESC-derived CMs [182]. Consequently, it is still unclear whether in either case the cells represented functional pacemaker cells.

Our own group combined overexpression of the highly conserved key nodal cell inducer Tbx3 with a neomycin resistance gene under the control of the well-established αMHC-promoter, leading to small aggregates consisting of ~300–500 cells, which we termed "induced sinoatrial bodies" (iSABs). The iSABS exhibited strongly increased beating frequencies of between 350 and 400 bpm in vitro, for the first time matching those of a murine heart. Extensive analyses, such as confocal laser scanning microscopy, FACS, single-cell patch clamping, funny channel density measuring, and Ca^{2+} imaging, revealed that the iSABs consisted of over 80% mature functional nodal cells with the rest representing immature nodal cells. To further address the pacemaker potential of iSABs, we relied on the ex vivo model of cultivated adult mouse ventricular slices. Thereby, the iSABs were capable of integrating into the slice tissue, remaining spontaneously active and pacing the slices to robust contractions. We confirmed their functional coupling to the slice via analysis of calcium transients, which proved to be synchronized between iSABs and slices. Therefore, iSABs represent highly pure murine PSC-derived nodal tissue, which is functional on the physiological level in vitro and in an ex vivo model [182]. The next step is to address the ability of iSABs to pace cardiac tissue in vivo. Likewise, the transferability of our approach to human PSCs has to be proven. Moreover, the knowledge gained from systems biology

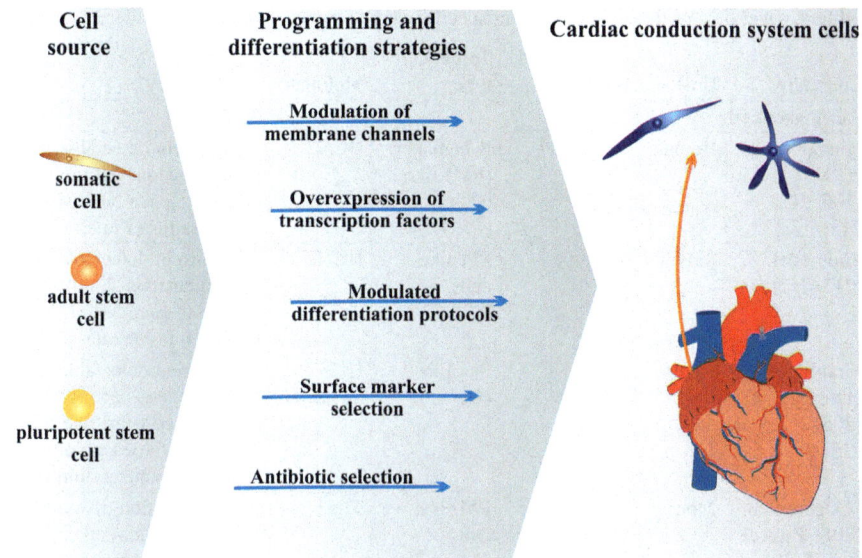

Fig. 3 Programming strategies for the generation of cardiac pacemaker cells. Somatic cells and different stem cell classes (*left*) are transformed or differentiated (*middle*) into cells belonging to the cardiac conduction system (*right*). Additional transgenic or nontransgenic selection strategies may be required (*middle*)

approaches such as RNAseq are crucial for further optimization of cell programming and purification (Fig. 3, Table 4) [232].

5   Conclusion

Reliable transdifferentiation of somatic cells or targeted differentiation of stem cells into highly pure mature cells offer exciting possibilities for regenerative medicine. Moreover, drug testing to treat metabolic disorders can benefit from this rapidly growing field. In addition, autologous cells from patients could enable deeper insights into molecular mechanisms underlying the disease.

Although current protocols do not yet suffice to yield fully functional cell types, such as cardiomyocytic subtypes, that are therapeutically or diagnostically usable, remarkable success in cellular (re-) programming over the last few years has significantly enhanced the field of regenerative medicine.

Table 4 Overview of the most prominent specific cardiac pacemaker creation strategies

Literature	Host	Original cell type	Target cell type	Modulator	Efficiency
Pacemaker cells					
Potapova et al. [206, 212, 213]	Human	MSC	Pacemaker (PM)-like cells	HCN2	I_f current detectable, no spontaneous activity, able to pace NRVMs with a faster rate
Zhou et al. [215–217]	Rabbit	MSC	PM-like cells	HCN2	I_f current detectable, no spontaneous activity, able to pace NRVMs with a faster rate
Plotnikov et al. [206, 212, 213]	Human	MSC	PM-like cells	HCN2	I_f current detectable, no spontaneous activity, increased heart rate for dogs with heart block after transplantation
Hoogars et al. [201], Bakker et al. [178]	Mouse	Working myocardium	PM-like cells	Tbx3	Upregulation of some PM-specific markers, no complete conversion into PM
Yang et al. [130, 207–210]	Rabbit	MSC	PM-like cells	HCN4	HCN4 detectable, no spontaneous activity, able to pace NRVMs with a faster rate
Tong et al. [130, 206–218]	Rat	MSC	PM-like cells	Cx45	I_f current detectable, no spontaneous activity, able to pace NRVMs with a faster rate
Kleger et al. [226]	Mouse	ESC	PM-like cells	EBIO	Upregulation of some PM-specific markers, frequency of ~160 bpm
Ma et al. [211, 218]	Rat	MSC	PM-like cells	HCN2	I_f current detectable, no spontaneous activity
Yang et al. [214, 219]	Human	BM-MSC, ATSC	PM-like cells	5-Azacytidin	PM-specific markers detectable, no spontaneous activity
Scavone et al. [228]	Mouse	ESC	PM-like cells	CD166+ (ALCAM) purification	Pacemaker characteristics, 10% HCN4+ cells
Lu et al. [130, 207–210]	Canine	MSC	PM-like cells	HCN4	I_f current detectable, no spontaneous activity, impulse generation shifted partially to injection site
Zhou et al. [207–209, 213, 216]	Rabbit	MSC	PM-like cells	HCN1	I_f current detectable, no spontaneous activity, able to pace NRVMs with a faster rate

(continued)

Table 4 (continued)

Literature	Host	Original cell type	Target cell type	Modulator	Efficiency
Hashem and Claycomb [230]	Mouse	ESC	PM-like cells	SHOX2	Upregulation of some PM-specific markers, frequency of ~120 bpm
Kapoor et al. [180], Hu et al. [179]	Rat, guinea pig, pig	NRVM, working myocardium	PM-like cells	TBX18	Increased beating frequencies, downregulation of myocardial markers, PM-like action potential, transient effect
Jung et al. [182] Rimmbach et al. [231]	Mouse	ESC	PM-like cells	TBX3 Myh6-promoter-based antibiotic selection	>80% physiologically and pharmacologically functional pacemaker cells with highly increased beating rates (300–400 bpm)
Ionta et al. [181]	Mouse	ESC	PM-like cells	SHOX2	Upregulation of some PM-specific markers, frequency of ~80 bpm
Feng et al. [130]	Canine	MSC	PM-like cells	HCN4, electric-pulse current stimulation	I_f current detectable, no spontaneous activity
Bruzauskaite et al. [206, 212, 213]	Human	MSC	PM-like cells	HCN2	I_f current detectable, HCN2 expression has no negative influence on cell viability
Feng et al. [130, 207, 208]	Canine	BM-MSC	PM-like cells	Shox2	I_f current detectable, no spontaneous activity, able to pace NRVMs with a faster rate
Chen et al. [219]	Mouse	ATSC	PM-like cells	Tbx18	Spontaneous beating cells out of brown adipose tissue SCs, decreasing pacemaker-specific markers by silencing Tbx18
Protze et al. [151]	Human	ESC, iPSC	PM-like cells	Specific culture conditions, SIRPA$^+$ purification	Pacemaker lineage-specific marker expression, increased beating frequencies, able to pace host tissue

References

1. Bergmann O, Bhardwaj RD, Bernard S, Zdunek S, Barnabe-Heider F, Walsh S et al (2009) Evidence for cardiomyocyte renewal in humans. Science 324:98–102
2. Mollova M, Bersell K, Walsh S, Savla J, Das LT, Park S-Y et al (2013) Cardiomyocyte proliferation contributes to heart growth in young humans. Proc Natl Acad Sci U S A 110:1446–1451
3. World Health Organization (WHO) The top 10 causes of death 2015. www.who.int/mediacentre/factsheets/fs310/en
4. Eurotransplant (2015) Annual Report 2015. Eurotransplant International Foundation, Leiden. Available at https://www.eurotransplant.org/cms/mediaobject.php?file=AR_ET_20153.pdf
5. Jain A, Bansal R (2015) Applications of regenerative medicine in organ transplantation. J Pharm Bioallied Sci 7:188–194
6. Heidary Rouchi A, Mahdavi-Mazdeh M (2015) Regenerative medicine in organ and tissue transplantation: shortly and practically achievable? Int J Organ Transplant Med 6:93–98
7. Orlando G, Soker S, Stratta RJ, Atala A (2013) Will regenerative medicine replace transplantation? Cold Spring Harb Perspect Med 3:a015693
8. Rosen MR, Myerburg RJ, Francis DP, Cole GD, Marbán E (2014) Translating stem cell research to cardiac disease therapies: pitfalls and prospects for improvement. J Am Coll Cardiol 64:922–937
9. Pavo N, Charwat S, Nyolczas N, Jakab A, Murlasits Z, Bergler-Klein J et al (2014) Cell therapy for human ischemic heart diseases: critical review and summary of the clinical experiences. J Mol Cell Cardiol 75:12–24
10. Matar AA, Chong JJ (2014) Stem cell therapy for cardiac dysfunction. SpringerPlus 3:440
11. de Feo D, Merlini A, Laterza C, Martino G (2012) Neural stem cell transplantation in central nervous system disorders: from cell replacement to neuroprotection. Curr Opin Neurol 25:322–333
12. Mothe AJ, Tator CH (2013) Review of transplantation of neural stem/progenitor cells for spinal cord injury. Int J Dev Neurosci 31:701–713
13. Stamm C, Kleine H-D, Choi Y-H, Dunkelmann S, Lauffs J-A, Lorenzen B et al (2007) Intramyocardial delivery of CD133+ bone marrow cells and coronary artery bypass grafting for chronic ischemic heart disease: safety and efficacy studies. J Thorac Cardiovasc Surg 133:717–725
14. Wang Z, Wang L, Su X, Pu J, Jiang M, He B (2017) Rational transplant timing and dose of mesenchymal stromal cells in patients with acute myocardial infarction: a meta-analysis of randomized controlled trials. Stem Cell Res Ther 8:21
15. Beck S, Lee B-K, Kim J (2015) Multi-layered global gene regulation in mouse embryonic stem cells. Cell Mol Life Sci 72:199–216
16. Molofsky AV, Pardal R, Morrison SJ (2004) Diverse mechanisms regulate stem cell self-renewal. Curr Opin Cell Biol 16:700–707
17. Odorico JS, Kaufman DS, Thomson JA (2001) Multilineage differentiation from human embryonic stem cell lines. Stem Cells (Dayton, Ohio) 19:193–204
18. Draper JS, Fox V (2003) Human embryonic stem cells: multilineage differentiation and mechanisms of self-renewal. Arch Med Res 34:558–564
19. Trounson A, DeWitt ND (2016) Pluripotent stem cells progressing to the clinic. Nat Rev Mol Cell Biol 17:194–200
20. Takahashi K, Yamanaka S (2006) Induction of pluripotent stem cells from mouse embryonic and adult fibroblast cultures by defined factors. Cell 126:663–676
21. Takahashi K, Tanabe K, Ohnuki M, Narita M, Ichisaka T, Tomoda K et al (2007) Induction of pluripotent stem cells from adult human fibroblasts by defined factors. Cell 131:861–872
22. Segers VFM, Lee RT (2008) Stem-cell therapy for cardiac disease. Nature 451:937–942
23. Okano H, Nakamura M, Yoshida K, Okada Y, Tsuji O, Nori S et al (2013) Steps toward safe cell therapy using induced pluripotent stem cells. Circ Res 112:523–533

24. Wesselschmidt RL (2011) The teratoma assay: an in vivo assessment of pluripotency. Methods Mol Biol (Clifton, NJ) 767:231–241
25. Nelakanti RV, Kooreman NG, Wu JC (2015) Teratoma formation: a tool for monitoring pluripotency in stem cell research. Curr Protoc Stem Cell Biol 32:4A.8.1–4A.817
26. Kamakura T, Makiyama T, Sasaki K, Yoshida Y, Wuriyanghai Y, Chen J et al (2013) Ultrastructural maturation of human-induced pluripotent stem cell-derived cardiomyocytes in a long-term culture. Circ J 77:1307–1314
27. Panopoulos AD, Ruiz S, Belmonte JCI (2011) iPSCs: induced back to controversy. Cell Stem Cell 8:347–348
28. Davis RL, Weintraub H, Lassar AB (1987) Expression of a single transfected cDNA converts fibroblasts to myoblasts. Cell Press 51:987–1000
29. Li X, Liu D, Ma Y, Du X, Jing J, Wang L et al (2017) Direct reprogramming of fibroblasts via a chemically induced XEN-like state. Cell Stem Cell. https://doi.org/10.1016/j.stem.2017.05.019
30. Heinrich C, Blum R, Gascón S, Masserdotti G, Tripathi P, Sánchez R et al (2010) Directing astroglia from the cerebral cortex into subtype specific functional neurons. PLoS Biol 8: e1000373
31. Caiazzo M, Dell'Anno MT, Dvoretskova E, Lazarevic D, Taverna S, Leo D et al (2011) Direct generation of functional dopaminergic neurons from mouse and human fibroblasts. Nature 476:224–227
32. Yoo AS, Sun AX, Li L, Shcheglovitov A, Portmann T, Li Y et al (2011) MicroRNA-mediated conversion of human fibroblasts to neurons. Nature 476:228–231
33. Guo Z, Zhang L, Wu Z, Chen Y, Wang F, Chen G (2014) In vivo direct reprogramming of reactive glial cells into functional neurons after brain injury and in an Alzheimer's disease model. Cell Stem Cell 14:188–202
34. Colasante G, Lignani G, Rubio A, Medrihan L, Yekhlef L, Sessa A et al (2015) Rapid conversion of fibroblasts into functional forebrain GABAergic interneurons by direct genetic reprogramming. Cell Stem Cell 17:719–734
35. Zhang L, Yin J-C, Yeh H, Ma N-X, Lee G, Chen XA et al (2015) Small molecules efficiently reprogram human astroglial cells into functional neurons. Cell Stem Cell 17:735–747
36. Li X, Zuo X, Jing J, Ma Y, Wang J, Liu D et al (2015) Small-molecule-driven direct reprogramming of mouse fibroblasts into functional neurons. Cell Stem Cell 17:195–203
37. Hu W, Qiu B, Guan W, Wang Q, Wang M, Li W et al (2015) Direct conversion of normal and alzheimer?: S disease human fibroblasts into neuronal cells by small molecules. Cell Stem Cell 17:204–212
38. He S, Guo Y, Zhang Y, Li Y, Feng C, Li X et al (2015) Reprogramming somatic cells to cells with neuronal characteristics by defined medium both in vitro and in vivo. Cell Regen (London, England) 4:12
39. Di Rivetti Val Cervo P, Romanov RA, Spigolon G, Masini D, Martín-Montañez E, Toledo EM et al (2017) Induction of functional dopamine neurons from human astrocytes in vitro and mouse astrocytes in a Parkinson's disease model. Nat Biotechnol 35:444–452
40. Huang P, He Z, Ji S, Sun H, Xiang D, Liu C et al (2011) Induction of functional hepatocyte-like cells from mouse fibroblasts by defined factors. Nature 475:386–389
41. Huang P, Zhang L, Gao Y, He Z, Yao D, Wu Z et al (2014) Direct reprogramming of human fibroblasts to functional and expandable hepatocytes. Cell Stem Cell 14:370–384
42. Du Y, Wang J, Jia J, Song N, Xiang C, Xu J et al (2014) Human hepatocytes with drug metabolic function induced from fibroblasts by lineage reprogramming. Cell Stem Cell 14:394–403
43. Simeonov KP, Uppal H (2014) Direct reprogramming of human fibroblasts to hepatocyte-like cells by synthetic modified mRNAs. PLoS One 9:e100134
44. Kim J, Kim K-P, Lim KT, Lee SC, Yoon J, Song G et al (2015) Generation of integration-free induced hepatocyte-like cells from mouse fibroblasts. Sci Rep 5:15706
45. Song G, Pacher M, Balakrishnan A, Yuan Q, Tsay H-C, Yang D et al (2016) Direct reprogramming of hepatic myofibroblasts into hepatocytes in vivo attenuates liver fibrosis. Cell Stem Cell 18:797–808

46. Rezvani M, Español-Suñer R, Malato Y, Dumont L, Grimm AA, Kienle E et al (2016) In vivo hepatic reprogramming of myofibroblasts with AAV vectors as a therapeutic strategy for liver fibrosis. Cell Stem Cell 18:809–816
47. Zhou Q, Brown J, Kanarek A, Rajagopal J, Melton DA (2008) In vivo reprogramming of adult pancreatic exocrine cells to beta-cells. Nature 455:627–632
48. Cavelti-Weder C, Li W, Weir GC, Zhou Q (2014) Direct lineage conversion of pancreatic exocrine to endocrine Beta cells in vivo with defined factors. Methods Mol Biol (Clifton, NJ) 1150:247–262
49. Banga A, Akinci E, Greder LV, Dutton JR, Slack JMW (2012) In vivo reprogramming of Sox9+ cells in the liver to insulin-secreting ducts. Proc Natl Acad Sci U S A 109:15336–15341
50. Lemper M, Leuckx G, Heremans Y, German MS, Heimberg H, Bouwens L et al (2015) Reprogramming of human pancreatic exocrine cells to β-like cells. Cell Death Differ 22:1117–1130
51. Zhu S, Russ HA, Wang X, Zhang M, Ma T, Xu T et al (2016) Human pancreatic beta-like cells converted from fibroblasts. Nat Commun 7:10080
52. Yang X-F, Ren L-W, Yang L, Deng C-Y, Li F-R (2017) In vivo direct reprogramming of liver cells to insulin producing cells by virus-free overexpression of defined factors. Endocr J 64:291–302
53. Yamamoto K, Kishida T, Sato Y, Nishioka K, Ejima A, Fujiwara H et al (2015) Direct conversion of human fibroblasts into functional osteoblasts by defined factors. Proc Natl Acad Sci U S A 112:6152–6157
54. Li Y, Wang Y, Yu J, Ma Z, Bai Q, Wu X et al (2017) Direct conversion of human fibroblasts into osteoblasts and osteocytes with small molecules and a single factor, Runx2. doi: 10.1101/127480
55. Warren L, Manos PD, Ahfeldt T, Loh Y-H, Li H, Lau F et al (2010) Highly efficient reprogramming to pluripotency and directed differentiation of human cells with synthetic modified mRNA. Cell Stem Cell 7:618–630
56. Hausburg F, Naß S, Voronina N, Skorska A, Müller P, Steinhoff G et al (2015) Defining optimized properties of modified mRNA to enhance virus- and DNA- independent protein expression in adult stem cells and fibroblasts. Cell Physiol Biochem 35:1360–1371
57. Bichsel C, Neeld D, Hamazaki T, Chang L-J, Yang L-J, Terada N et al (2013) Direct reprogramming of fibroblasts to myocytes via bacterial injection of MyoD protein. Cell Reprogram 15:117–125
58. Kim EY, Page P, Dellefave-Castillo LM, McNally EM, Wyatt EJ (2016) Direct reprogramming of urine-derived cells with inducible MyoD for modeling human muscle disease. Skelet Muscle 6:32
59. Horio F, Sakurai H, Ohsawa Y, Nakano S, Matsukura M, Fujii I (2017) Functional validation and expression analysis of myotubes converted from skin fibroblasts using a simple direct reprogramming strategy. eNeurologicalSci 6:9–15
60. Ieda M, Fu J-D, Delgado-Olguin P, Vedantham V, Hayashi Y, Bruneau BG et al (2010) Direct reprogramming of fibroblasts into functional cardiomyocytes by defined factors. Cell 142:375–386
61. Chen JX, Krane M, Deutsch M-A, Wang L, Rav-Acha M, Gregoire S et al (2012) Inefficient reprogramming of fibroblasts into cardiomyocytes using Gata4, Mef2c, and Tbx5. Circ Res 111:50–55
62. Qian L, Huang Y, Spencer CI, Foley A, Vedantham V, Liu L et al (2012) In vivo reprogramming of murine cardiac fibroblasts into induced cardiomyocytes. Nature 485:593–598
63. Inagawa K, Miyamoto K, Yamakawa H, Muraoka N, Sadahiro T, Umei T et al (2012) Induction of cardiomyocyte-like cells in infarct hearts by gene transfer of Gata4, Mef2c, and Tbx5. Circ Res 111:1147–1156

64. Qian L, Berry EC, Fu J-D, Ieda M, Srivastava D (2013) Reprogramming of mouse fibroblasts into cardiomyocyte-like cells in vitro. Nat Protoc 8:1204–1215
65. Wang L, Liu Z, Yin C, Asfour H, Chen O, Li Y et al (2015) Stoichiometry of Gata4, Mef2c, and Tbx5 influences the efficiency and quality of induced cardiac myocyte reprogramming. Circ Res 116:237–244
66. Song K, Nam Y-J, Luo X, Qi X, Tan W, Huang GN et al (2012) Heart repair by reprogramming non-myocytes with cardiac transcription factors. Nature 485:599–604
67. Jayawardena TM, Egemnazarov B, Finch EA, Zhang L, Payne JA, Pandya K et al (2012) MicroRNA-mediated in vitro and in vivo direct reprogramming of cardiac fibroblasts to cardiomyocytes. Circ Res 110:1465–1473
68. Jayawardena T, Mirotsou M, Dzau VJ (2014) Direct reprogramming of cardiac fibroblasts to cardiomyocytes using microRNAs. Methods Mol Biol 1150:263–272
69. Jayawardena TM, Finch EA, Zhang L, Zhang H, Hodgkinson CP, Pratt RE et al (2015) MicroRNA induced cardiac reprogramming in vivo: evidence for mature cardiac myocytes and improved cardiac function. Circ Res 116:418–424
70. Nam Y-J, Song K, Luo X, Daniel E, Lambeth K, West K et al (2013) Reprogramming of human fibroblasts toward a cardiac fate. Proc Natl Acad Sci U S A 110:5588–5593
71. Hirai H, Katoku-Kikyo N, Keirstead SA, Kikyo N (2013) Accelerated direct reprogramming of fibroblasts into cardiomyocyte-like cells with the MyoD transactivation domain. Cardiovasc Res 100:105–113
72. Hirai H, Kikyo N (2014) Inhibitors of suppressive histone modification promote direct reprogramming of fibroblasts to cardiomyocyte-like cells. Cardiovasc Res 102:188–190
73. Wang H, Cao N, Spencer CI, Nie B, Ma T, Xu T et al (2014) Small molecules enable cardiac reprogramming of mouse fibroblasts with a single factor, Oct4. Cell Rep 6:951–960
74. Fu Y, Huang C, Xu X, Gu H, Ye Y, Jiang C et al (2015) Direct reprogramming of mouse fibroblasts into cardiomyocytes with chemical cocktails. Cell Res 25:1013–1024
75. Zhao Y, Londono P, Cao Y, Sharpe EJ, Proenza C, O'Rourke R et al (2015) High-efficiency reprogramming of fibroblasts into cardiomyocytes requires suppression of pro-fibrotic signalling. Nat Commun 6:8243
76. Zhou H, Dickson ME, Kim MS, Bassel-Duby R, Olson EN (2015) Akt1/protein kinase B enhances transcriptional reprogramming of fibroblasts to functional cardiomyocytes. Proc Natl Acad Sci U S A 112:11864–11869
77. Yamakawa H, Muraoka N, Miyamoto K, Sadahiro T, Isomi M, Haginiwa S et al (2015) Fibroblast growth factors and vascular endothelial growth factor promote cardiac reprogramming under defined conditions. Stem Cell Rep 5:1128–1142
78. Talkhabi M, Pahlavan S, Aghdami N, Baharvand H (2015) Ascorbic acid promotes the direct conversion of mouse fibroblasts into beating cardiomyocytes. Biochem Biophys Res Commun 463:699–705
79. Park G, Yoon BS, Kim YS, Choi S-C, Moon J-H, Kwon S et al (2015) Conversion of mouse fibroblasts into cardiomyocyte-like cells using small molecule treatments. Biomaterials 54:201–212
80. Cao N, Huang Y, Zheng J, Spencer CI, Zhang Y, Fu J-D et al (2016) Conversion of human fibroblasts into functional cardiomyocytes by small molecules. Science (New York, NY) 352:1216–1220
81. Zhou Y, Wang L, Vaseghi HR, Liu Z, Lu R, Alimohamadi S et al (2016) Bmi1 is a key epigenetic barrier to direct cardiac reprogramming. Cell Stem Cell 18:382–395
82. Mohamed TMA, Stone NR, Berry EC, Radzinsky E, Huang Y, Pratt K et al (2017) Chemical enhancement of in vitro and in vivo direct cardiac reprogramming. Circulation 135:978–995
83. Hughes RD, Mitry RR, Dhawan A (2012) Current status of hepatocyte transplantation. Transplantation 93:342–347
84. Laflamme MA, Murry CE (2011) Heart regeneration. Nature 473:326–335
85. Bhatia SK (2010) Biomaterials for clinical applications

86. Whelan RS, Kaplinskiy V, Kitsis RN (2010) Cell death in the pathogenesis of heart disease: mechanisms and significance. Annu Rev Physiol 72:19–44
87. Anversa P, Kajstura J (1998) Myocyte cell death in the diseased heart. Circ Res 82:1231–1233
88. Gottlieb RA, Burleson KO, Kloner RA, Babior BM, Engler RL (1994) Reperfusion injury induces apoptosis in rabbit cardiomyocytes. J Clin Invest 94:1621–1628
89. Xia P, Liu Y, Cheng Z (2016) Signaling pathways in cardiac myocyte apoptosis. Biomed Res Int 2016:9583268
90. Baines CP, Kaiser RA, Purcell NH, Blair NS, Osinska H, Hambleton MA et al (2005) Loss of cyclophilin D reveals a critical role for mitochondrial permeability transition in cell death. Nature 434:658–662
91. Nakagawa T, Shimizu S, Watanabe T, Yamaguchi O, Otsu K, Yamagata H et al (2005) Cyclophilin D-dependent mitochondrial permeability transition regulates some necrotic but not apoptotic cell death. Nature 434:652–658
92. Jennings RB, Sommers HM, Smyth GA, Flack HA, Linn H (1960) Myocardial necrosis induced by temporary occlusion of a coronary artery in the dog. Arch Pathol 70:68–78
93. Matsui Y, Takagi H, Qu X, Abdellatif M, Sakoda H, Asano T et al (2007) Distinct roles of autophagy in the heart during ischemia and reperfusion: roles of AMP-activated protein kinase and Beclin 1 in mediating autophagy. Circ Res 100:914–922
94. Takagi H, Matsui Y, Hirotani S, Sakoda H, Asano T, Sadoshima J (2007) AMPK mediates autophagy during myocardial ischemia in vivo. Autophagy 3:405–407
95. Hamacher-Brady A, Brady NR, Gottlieb RA (2006) Enhancing macroautophagy protects against ischemia/reperfusion injury in cardiac myocytes. J Biol Chem 281:29776–29787
96. Murry CE, Reinecke H, Pabon LM (2006) Regeneration gaps: observations on stem cells and cardiac repair. J Am Coll Cardiol 47:1777–1785
97. Laflamme MA, Murry CE (2005) Regenerating the heart. Nat Biotechnol 23:845–856
98. Hasenfuss G (1998) Animal models of human cardiovascular disease, heart failure and hypertrophy. Cardiovasc Res 39:60–76
99. Frangogiannis NG (2015) Pathophysiology of myocardial infarction. Compr Physiol 5:1841–1875
100. Richardson WJ, Clarke SA, Quinn TA, Holmes JW (2015) Physiological implications of myocardial scar structure. Compr Physiol 5:1877–1909
101. Leor J, Palevski D, Amit U, Konfino T (2016) Macrophages and regeneration: lessons from the heart. Semin Cell Dev Biol 58:26–33
102. Ramos G, Hofmann U, Frantz S (2016) Myocardial fibrosis seen through the lenses of T-cell biology. J Mol Cell Cardiol 92:41–45
103. Lighthouse JK, Small EM (2016) Transcriptional control of cardiac fibroblast plasticity. J Mol Cell Cardiol 91:52–60
104. Saez P, Kuhl E (2016) Computational modeling of acute myocardial infarction. Comput Methods Biomech Biomed Eng 19:1107–1115
105. Cheng B, Chen HC, Chou IW, Tang TWH, Hsieh PCH (2017) Harnessing the early post-injury inflammatory responses for cardiac regeneration. J Biomed Sci 24:7
106. Ghosh AK, Rai R, Flevaris P, Vaughan DE (2017) Epigenetics in reactive and reparative cardiac fibrogenesis: the promise of epigenetic therapy. J Cell Physiol 232:1941–1956. doi:10.1002/jcp.25699
107. Turner NA (2016) Inflammatory and fibrotic responses of cardiac fibroblasts to myocardial damage associated molecular patterns (DAMPs). J Mol Cell Cardiol 94:189–200
108. Chen B, Frangogiannis NG (2017) Immune cells in repair of the infarcted myocardium. Microcirculation 24:e12305. https://doi.org/10.1111/micc.12305
109. Chistiakov DA, Orekhov AN, Bobryshev YV (2016) The role of cardiac fibroblasts in post-myocardial heart tissue repair. Exp Mol Pathol 101:231–240
110. Kurose H, Mangmool S (2016) Myofibroblasts and inflammatory cells as players of cardiac fibrosis. Arch Pharm Res 39:1100–1113

111. Garbern JC, Lee RT (2013) Cardiac stem cell therapy and the promise of heart regeneration. Cell Stem Cell 12:689–698
112. Senyo SE, Steinhauser ML, Pizzimenti CL, Yang VK, Cai L, Wang M et al (2013) Mammalian heart renewal by pre-existing cardiomyocytes. Nature 493:433–436
113. Gao Y, Chu M, Hong J, Shang J, Di X (2014) Hypoxia induces cardiac fibroblast proliferation and phenotypic switch: a role for caveolae and caveolin-1/PTEN mediated pathway. J Thorac Dis 6:1458–1468
114. Moore-Morris T, Cattaneo P, Puceat M, Evans SM (2016) Origins of cardiac fibroblasts. J Mol Cell Cardiol 91:1–5
115. Kamps JA, Krenning G (2016) Micromanaging cardiac regeneration: targeted delivery of microRNAs for cardiac repair and regeneration. World J Cardiol 8:163–179
116. Fu J-D, Rushing SN, Lieu DK, Chan CW, Kong C-W, Geng L et al (2011) Distinct roles of microRNA-1 and -499 in ventricular specification and functional maturation of human embryonic stem cell-derived cardiomyocytes. PLoS One 6:e27417
117. Wilson KD, Hu S, Venkatasubrahmanyam S, Fu J-D, Sun N, Abilez OJ et al (2010) Dynamic microRNA expression programs during cardiac differentiation of human embryonic stem cells: role for miR-499. Circ Cardiovasc Genet 3:426–435
118. Zhao Y, Samal E, Srivastava D (2005) Serum response factor regulates a muscle-specific microRNA that targets Hand2 during cardiogenesis. Nature 436:214–220
119. Liu N, Williams AH, Kim Y, McAnally J, Bezprozvannaya S, Sutherland LB et al (2007) An intragenic MEF2-dependent enhancer directs muscle-specific expression of microRNAs 1 and 133. Proc Natl Acad Sci U S A 104:20844–20849
120. Qian L, Wythe JD, Liu J, Cartry J, Vogler G, Mohapatra B et al (2011) Tinman/Nkx2-5 acts via miR-1 and upstream of Cdc42 to regulate heart function across species. J Cell Biol 193:1181–1196
121. Ivey KN, Muth A, Arnold J, King FW, Yeh R-F, Fish JE et al (2008) MicroRNA regulation of cell lineages in mouse and human embryonic stem cells. Cell Stem Cell 2:219–229
122. Huang F, Tang L, Fang Z-f, Hu X-q, Pan J-y, Zhou S-h (2013) miR-1-mediated induction of cardiogenesis in mesenchymal stem cells via downregulation of Hes-1. BioMed Res Int 2013:216286
123. Nagalingam RS, Safi HA, Czubryt MP (2016) Gaining myocytes or losing fibroblasts: challenges in cardiac fibroblast reprogramming for infarct repair. J Mol Cell Cardiol 93:108–114
124. Andrée B, Zweigerdt R (2016) Directing Cardiomyogenic differentiation and Transdifferentiation by ectopic gene expression - direct transition or reprogramming detour? CGT 16:14–20
125. Karpov AA, Udalova DV, Pliss MG, Galagudza MM (2016) Can the outcomes of mesenchymal stem cell-based therapy for myocardial infarction be improved? Providing weapons and armour to cells. Cell Prolif 50(2):e12316. doi:10.1111/cpr.12316
126. Yang W, Zheng H, Wang Y, Lian F, Hu Z, Xue S (2015) Nesprin-1 has key roles in the process of mesenchymal stem cell differentiation into cardiomyocyte-like cells in vivo and in vitro. Mol Med Rep 11:133–142
127. Li P, Zhang L (2015) Exogenous Nkx2.5- or GATA-4-transfected rabbit bone marrow mesenchymal stem cells and myocardial cell co-culture on the treatment of myocardial infarction in rabbits. Mol Med Rep 12:2607–2621
128. Li J, Zhu K, Wang Y, Zheng J, Guo C, Lai H et al (2015) Combination of IGF1 gene manipulation and 5AZA treatment promotes differentiation of mesenchymal stem cells into cardiomyocyte-like cells. Mol Med Rep 11:815–820
129. Mohanty S, Bose S, Jain KG, Bhargava B, Airan B (2013) TGFβ1 contributes to cardiomyogenic-like differentiation of human bone marrow mesenchymal stem cells. Int J Cardiol 163:93–99
130. Feng Y, Yang P, Luo S, Zhang Z, Li H, Zhu P et al (2016) Shox2 influences mesenchymal stem cell fate in a co-culture model in vitro. Mol Med Rep 14:637–642

131. Yu Z, Zou Y, Fan J, Li C, Ma L (2016) Notch1 is associated with the differentiation of human bone marrow-derived mesenchymal stem cells to cardiomyocytes. Mol Med Rep 14:5065–5071

132. Hou J, Long H, Zhou C, Zheng S, Wu H, Guo T et al (2017) Long noncoding RNA Braveheart promotes cardiogenic differentiation of mesenchymal stem cells in vitro. Stem Cell Res Ther 8:4

133. Carvalho PH, Daibert APF, Monteiro BS, Okano BS, Carvalho JL, da Cunha DNQ et al (2013) Diferenciação de células-tronco mesenquimais derivadas do tecido adiposo em cardiomiócitos. Arq Bras Cardiol 100:82–89

134. Wystrychowski W, Patlolla B, Zhuge Y, Neofytou E, Robbins RC, Beygui RE (2016) Multipotency and cardiomyogenic potential of human adipose-derived stem cells from epicardium, pericardium, and omentum. Stem Cell Res Ther 7:84

135. Gwak S-J, Bhang SH, Yang HS, Kim S-S, Lee D-H, Lee S-H et al (2009) In vitro cardiomyogenic differentiation of adipose-derived stromal cells using transforming growth factor-beta1. Cell Biochem Funct 27:148–154

136. Nagata H, Ii M, Kohbayashi E, Hoshiga M, Hanafusa T, Asahi M (2016) Cardiac adipose-derived stem cells exhibit high differentiation potential to cardiovascular cells in C57BL/6 mice. Stem Cells Transl Med 5:141–151

137. Choi YS, Dusting GJ, Stubbs S, Arunothayaraj S, Han XL, Collas P et al (2010) Differentiation of human adipose-derived stem cells into beating cardiomyocytes. J Cell Mol Med 14:878–889

138. Takahashi T, Nagai T, Kanda M, Liu M-L, Kondo N, Naito AT et al (2015) Regeneration of the cardiac conduction system by adipose tissue-derived stem cells. Circ J 79:2703–2712

139. Sung I-Y, Son H-N, Ullah I, Bharti D, Park J-M, Cho Y-C et al (2016) Cardiomyogenic differentiation of human dental follicle-derived stem cells by suberoylanilide hydroxamic acid and their in vivo homing property. Int J Med Sci 13:841–852

140. Lopez-Ruiz E, Peran M, Picon-Ruiz M, Garcia MA, Carrillo E, Jimenez-Navarro M et al (2014) Cardiomyogenic differentiation potential of human endothelial progenitor cells isolated from patients with myocardial infarction. Cytotherapy 16:1229–1237

141. Hosoda T, Zheng H, Cabral-da-Silva M, Sanada F, Ide-Iwata N, Ogórek B et al (2011) Human cardiac stem cell differentiation is regulated by a mircrine mechanism. Circulation 123:1287–1296

142. Goumans M-J, de Boer TP, Smits AM, van Laake LW, van Vliet P, Metz CHG et al (2007) TGF-beta1 induces efficient differentiation of human cardiomyocyte progenitor cells into functional cardiomyocytes in vitro. Stem Cell Res 1:138–149

143. Sluijter JPG, van Mil A, van Vliet P, Metz CHG, Liu J, Doevendans PA et al (2010) MicroRNA-1 and -499 regulate differentiation and proliferation in human-derived cardiomyocyte progenitor cells. Arterioscler Thromb Vasc Biol 30:859–868

144. Avitabile D, Crespi A, Brioschi C, Parente V, Toietta G, Devanna P et al (2011) Human cord blood CD34+ progenitor cells acquire functional cardiac properties through a cell fusion process. Am J Phys Heart Circ Phys 300:H1875–H1884

145. Freeman BT, Kouris NA, Ogle BM (2015) Tracking fusion of human mesenchymal stem cells after transplantation to the heart. Stem Cells Transl Med 4:685–694

146. Kempf H, Zweigerdt R (2017) Scalable cardiac differentiation of pluripotent stem cells using specific growth factors and small molecules. Adv Biochem Eng/Biotechnol. https://doi.org/10.1007/10_2017_XX

147. Yang X, Pabon L, Murry CE (2014) Engineering adolescence: maturation of human pluripotent stem cell-derived cardiomyocytes. Circ Res 114:511–523

148. Veerman CC, Kosmidis G, Mummery CL, Casini S, Verkerk AO, Bellin M (2015) Immaturity of human stem-cell-derived cardiomyocytes in culture: fatal flaw or soluble problem? Stem Cells Dev 24:1035–1052

149. Bhattacharya S, Burridge PW, Kropp EM, Chuppa SL, Kwok W-M, Wu JC et al (2014) High efficiency differentiation of human pluripotent stem cells to cardiomyocytes and characterization by flow cytometry. J Vis Exp 91:52010

150. Chen VC, Ye J, Shukla P, Hua G, Chen D, Lin Z et al (2015) Development of a scalable suspension culture for cardiac differentiation from human pluripotent stem cells. Stem Cell Res 15:365–375
151. Protze SI, Liu J, Nussinovitch U, Ohana L, Backx PH, Gepstein L et al (2017) Sinoatrial node cardiomyocytes derived from human pluripotent cells function as a biological pacemaker. Nat Biotechnol 35:56–68
152. Pei F, Jiang J, Bai S, Cao H, Tian L, Zhao Y et al (2017) Chemical-defined and albumin-free generation of human atrial and ventricular myocytes from human pluripotent stem cells. Stem Cell Res 19:94–103
153. Fuerstenau-Sharp M, Zimmermann ME, Stark K, Jentsch N, Klingenstein M, Drzymalski M et al (2015) Generation of highly purified human cardiomyocytes from peripheral blood mononuclear cell-derived induced pluripotent stem cells. PLoS One 10:e0126596
154. Mazzotta S, Neves C, Bonner RJ, Bernardo AS, Docherty K, Hoppler S (2016) Distinctive roles of canonical and noncanonical Wnt signaling in human embryonic cardiomyocyte development. Stem Cell Rep 7:764–776
155. Kempf H, Olmer R, Kropp C, Rückert M, Jara-Avaca M, Robles-Diaz D et al (2014) Controlling expansion and cardiomyogenic differentiation of human pluripotent stem cells in scalable suspension culture. Stem Cell Rep 3:1132–1146
156. Kempf H, Olmer R, Haase A, Franke A, Bolesani E, Schwanke K et al (2016) Bulk cell density and Wnt/TGFbeta signalling regulate mesendodermal patterning of human pluripotent stem cells. Nat Commun 7:13602
157. Chen VC, Couture SM, Ye J, Lin Z, Hua G, Huang H-IP et al (2012) Scalable GMP compliant suspension culture system for human ES cells. Stem Cell Res 8:388–402
158. Palpant NJ, Pabon L, Friedman CE, Roberts M, Hadland B, Zaunbrecher RJ et al (2017) Generating high-purity cardiac and endothelial derivatives from patterned mesoderm using human pluripotent stem cells. Nat Protoc 12:15–31
159. Hattori F, Chen H, Yamashita H, Tohyama S, Satoh Y-S, Yuasa S et al (2010) Nongenetic method for purifying stem cell-derived cardiomyocytes. Nat Methods 7:61–66
160. Dubois NC, Craft AM, Sharma P, Elliott DA, Stanley EG, Elefanty AG et al (2011) SIRPA is a specific cell-surface marker for isolating cardiomyocytes derived from human pluripotent stem cells. Nat Biotechnol 29:1011–1018
161. van Hoof D, Dormeyer W, Braam SR, Passier R, Monshouwer-Kloots J, Ward-van Oostwaard D et al (2010) Identification of cell surface proteins for antibody-based selection of human embryonic stem cell-derived cardiomyocytes. J Proteome Res 9:1610–1618
162. Uosaki H, Fukushima H, Takeuchi A, Matsuoka S, Nakatsuji N, Yamanaka S et al (2011) Efficient and scalable purification of cardiomyocytes from human embryonic and induced pluripotent stem cells by VCAM1 surface expression. PLoS One 6:e23657
163. Tohyama S, Hattori F, Sano M, Hishiki T, Nagahata Y, Matsuura T et al (2013) Distinct metabolic flow enables large-scale purification of mouse and human pluripotent stem cell-derived cardiomyocytes. Cell Stem Cell 12:127–137
164. Zhang J, Wilson GF, Soerens AG, Koonce CH, Yu J, Palecek SP et al (2009) Functional cardiomyocytes derived from human induced pluripotent stem cells. Circ Res 104:e30–e41
165. Pasquier J, Gupta R, Rioult D, Hoarau-Vechot J, Courjaret R, Machaca K et al (2017) Coculturing with endothelial cells promotes in vitro maturation and electrical coupling of human embryonic stem cell-derived cardiomyocytes. J Heart Lung Transplant 36(6):684–693
166. Kensah G, Roa Lara A, Dahlmann J, Zweigerdt R, Schwanke K, Hegermann J et al (2013) Murine and human pluripotent stem cell-derived cardiac bodies form contractile myocardial tissue in vitro. Eur Heart J 34:1134–1146
167. Oberwallner B, Brodarac A, Anic P, Saric T, Wassilew K, Neef K et al (2015) Human cardiac extracellular matrix supports myocardial lineage commitment of pluripotent stem cells. Eur J Cardiothorac Surg 47:416–425. discussion 425
168. Amano Y, Nishiguchi A, Matsusaki M, Iseoka H, Miyagawa S, Sawa Y et al (2016) Development of vascularized iPSC derived 3D-cardiomyocyte tissues by filtration

layer-by-layer technique and their application for pharmaceutical assays. Acta Biomater 33:110–121

169. Valarmathi MT, Fuseler JW, Davis JM, Price RL (2017) A novel human tissue-engineered 3-D functional vascularized cardiac muscle construct. Front Cell Dev Biol 5:2

170. Eder A, Vollert I, Hansen A, Eschenhagen T (2016) Human engineered heart tissue as a model system for drug testing. Adv Drug Deliv Rev 96:214–224

171. Kempf H, Andree B, Zweigerdt R (2016) Large-scale production of human pluripotent stem cell derived cardiomyocytes. Adv Drug Deliv Rev 96:18–30

172. Walsh-Irwin C, Hannibal GB (2015) Sick sinus syndrome. AACN Adv Crit Care 26:376–380

173. Dobrzynski H, Boyett MR, Anderson RH (2007) New insights into pacemaker activity: promoting understanding of sick sinus syndrome. Circulation 115:1921–1932

174. Ewy GA (2014) Sick sinus syndrome: synopsis. J Am Coll Cardiol 64:539–540

175. Semelka M, Gera J, Usman S (2013) Sick sinus syndrome: a review. Am Fam Physician 87:691–696

176. Gregoratos G (2005) Indications and recommendations for pacemaker therapy. Am Fam Physician 71:1563–1570

177. Tse G, Liu T, Li KH, Laxton V, Wong AO, Chan YW et al (2017) Tachycardia-bradycardia syndrome: electrophysiological mechanisms and future therapeutic approaches (review). Int J Mol Med 39:519–526

178. Bakker ML, Boink GJ, Boukens BJ, Verkerk AO, van den Boogaard M, den Haan AD et al (2012) T-box transcription factor TBX3 reprogrammes mature cardiac myocytes into pacemaker-like cells. Cardiovasc Res 94:439–449

179. Hu Y-F, Dawkins JF, Cho HC, Marban E, Cingolani E (2014) Biological pacemaker created by minimally invasive somatic reprogramming in pigs with complete heart block. Sci Transl Med 6:245ra94

180. Kapoor N, Liang W, Marban E, Cho HC (2013) Direct conversion of quiescent cardiomyocytes to pacemaker cells by expression of Tbx18. Nat Biotechnol 31:54–62

181. Ionta V, Liang W, Kim EH, Rafie R, Giacomello A, Marban E et al (2015) SHOX2 overexpression favors differentiation of embryonic stem cells into cardiac pacemaker cells, improving biological pacing ability. Stem Cell Rep 4:129–142

182. Jung JJ, Husse B, Rimmbach C, Krebs S, Stieber J, Steinhoff G et al (2014) Programming and isolation of highly pure physiologically and pharmacologically functional sinus-nodal bodies from pluripotent stem cells. Stem Cell Rep 2:592–605

183. Lown B (1967) Electrical reversion of cardiac arrhythmias. Br Heart J 29:469–489

184. Johns DC, Nuss HB, Chiamvimonvat N, Ramza BM, Marban E, Lawrence JH (1995) Adenovirus-mediated expression of a voltage-gated potassium channel in vitro (rat cardiac myocytes) and in vivo (rat liver). A novel strategy for modifying excitability. J Clin Invest 96:1152–1158

185. Miake J, Marbán E, Nuss HB (2002) Biological pacemaker created by gene transfer. Nature 419:132–133

186. Schram G (2002) Differential distribution of cardiac ion channel expression as a basis for regional specialization in electrical function. Circ Res 90:939–950

187. Kubo Y, Baldwin TJ, Jan YN, Jan LY (1993) Primary structure and functional expression of a mouse inward rectifier potassium channel. Nature 362:127–133

188. Tinker A, Jan YN, Jan LY (1996) Regions responsible for the assembly of inwardly rectifying potassium channels. Cell 87:857–868

189. Miake J, Marbán E, Nuss HB (2003) Functional role of inward rectifier current in heart probed by Kir2.1 overexpression and dominant-negative suppression. J Clin Invest 111:1529–1536

190. Brugada P, Wellens HJ (1985) Early afterdepolarizations: role in conduction block, "prolonged repolarization-dependent reexcitation," and tachyarrhythmias in the human heart. Pacing Clin Electrophysiol 8:889–896

191. January CT, Moscucci A (1992) Cellular mechanisms of early afterdepolarizations. Ann N Y Acad Sci 644:23–32

192. Ennis IL, Li RA, Murphy AM, Marbán E, Nuss HB (2002) Dual gene therapy with SERCA1 and Kir2.1 abbreviates excitation without suppressing contractility. J Clin Invest 109:393–400
193. Baruscotti M, Bucchi A, Difrancesco D (2005) Physiology and pharmacology of the cardiac pacemaker ("funny") current. Pharmacol Ther 107:59–79
194. Plotnikov AN, Sosunov EA, Qu J, Shlapakova IN, Anyukhovsky EP, Liu L et al (2004) Biological pacemaker implanted in canine left bundle branch provides ventricular escape rhythms that have physiologically acceptable rates. Circulation 109:506–512
195. Qu J, Plotnikov A, Danilo P, Shlapakova I, Cohen IS, Robinson RB et al (2003) Expression and function of a biological pacemaker in canine heart. Circulation 107:1106–1109
196. Tse H-F, Xue T, Lau C-P, Siu C-W, Wang K, Zhang Q-Y et al (2006) Bioartificial sinus node constructed via in vivo gene transfer of an engineered pacemaker HCN Channel reduces the dependence on electronic pacemaker in a sick-sinus syndrome model. Circulation 114:1000–1011
197. Kashiwakura Y, Cho HC, Barth AS, Azene E, Marbán E (2006) Gene transfer of a synthetic pacemaker channel into the heart: a novel strategy for biological pacing. Circulation 114:1682–1686
198. Cho HC, Kashiwakura Y, Marbán E (2007) Creation of a biological pacemaker by cell fusion. Circ Res 100:1112–1115
199. Dorn T, Goedel A, Lam JT, Haas J, Tian Q, Herrmann F et al (2015) Direct nkx2-5 transcriptional repression of isl1 controls cardiomyocyte subtype identity. Stem Cells (Dayton, Ohio) 33:1113–1129
200. Frank DU, Carter KL, Thomas KR, Burr RM, Bakker ML, Coetzee WA et al (2012) Lethal arrhythmias in Tbx3-deficient mice reveal extreme dosage sensitivity of cardiac conduction system function and homeostasis. Proc Natl Acad Sci U S A 109:E154–E163
201. Hoogaars WMH, Engel A, Brons JF, Verkerk AO, de Lange FJ, Wong LYE et al (2007) Tbx3 controls the sinoatrial node gene program and imposes pacemaker function on the atria. Genes Dev 21:1098–1112
202. Vedantham V, Galang G, Evangelista M, Deo RC, Srivastava D (2015) RNA sequencing of mouse sinoatrial node reveals an upstream regulatory role for islet-1 in cardiac pacemaker cells. Circ Res 116:797–803
203. Wiese C, Grieskamp T, Airik R, Mommersteeg MTM, Gardiwal A, de Gier-de Vrie C et al (2009) Formation of the sinus node head and differentiation of sinus node myocardium are independently regulated by Tbx18 and Tbx3. Circ Res 104:388–397
204. Greulich F, Trowe M-O, Leffler A, Stoetzer C, Farin HF, Kispert A (2016) Misexpression of Tbx18 in cardiac chambers of fetal mice interferes with chamber-specific developmental programs but does not induce a pacemaker-like gene signature. J Mol Cell Cardiol 97:140–149
205. Nam Y-J, Lubczyk C, Bhakta M, Zang T, Fernandez-Perez A, McAnally J et al (2014) Induction of diverse cardiac cell types by reprogramming fibroblasts with cardiac transcription factors. Development 141:4267–4278
206. Bruzauskaite I, Bironaite D, Bagdonas E, Skeberdis VA, Denkovskij J, Tamulevicius T et al (2016) Relevance of HCN2-expressing human mesenchymal stem cells for the generation of biological pacemakers. Stem Cell Res Ther 7:67
207. Feng Y, Luo S, Tong S, Zhong L, Zhang C, Yang P et al (2015) Electric-pulse current stimulation increases if current in mShox2 genetically modified canine mesenchymal stem cells. Cardiology 132:49–57
208. Feng Y, Luo S, Yang P, Song Z (2016) Electric pulse current stimulation increases electrophysiological properties of I_f current reconstructed in mHCN4-transfected canine mesenchymal stem cells. Exp Ther Med 11:1323–1329
209. Jun C, Zhihui Z, Lu W, Yaoming N, Lei W, Yao Q et al (2012) Canine bone marrow mesenchymal stromal cells with lentiviral mHCN4 gene transfer create cardiac pacemakers. Cytotherapy 14:529–539

210. Lu W, Yaoming N, Boli R, Jun C, Changhai Z, Yang Z et al (2013) mHCN4 genetically modified canine mesenchymal stem cells provide biological pacemaking function in complete dogs with atrioventricular block. Pacing Clin Electrophysiol 36:1138–1149

211. Ma J, Zhang C, Huang S, Wang G, Quan X (2010) Use of rats mesenchymal stem cells modified with mHCN2 gene to create biologic pacemakers. J Huazhong Univ Sci Technol Med Sci 30:447–452

212. Plotnikov AN, Shlapakova I, Szabolcs MJ, Danilo P, Lorell BH, Potapova IA et al (2007) Xenografted adult human mesenchymal stem cells provide a platform for sustained biological pacemaker function in canine heart. Circulation 116:706–713

213. Potapova I, Plotnikov A, Lu Z, Danilo P, Valiunas V, Qu J et al (2004) Human mesenchymal stem cells as a gene delivery system to create cardiac pacemakers. Circ Res 94:952–959

214. Yang J, Song T, Wu P, Chen Y, Fan X, Chen H et al (2012) Differentiation potential of human mesenchymal stem cells derived from adipose tissue and bone marrow to sinus node-like cells. Mol Med Rep 5:108–113

215. Yang X-J, Zhou Y-F, Li H-X, Han L-H, Jiang W-P (2008) Mesenchymal stem cells as a gene delivery system to create biological pacemaker cells in vitro. J Int Med Res 36:1049–1055

216. Zhou Y-F, Yang X-J, Li H-X, Han L-H, Jiang W-P (2013) Genetically-engineered mesenchymal stem cells transfected with human HCN1 gene to create cardiac pacemaker cells. J Int Med Res 41:1570–1576

217. Zhou Y-F, Yang X-J, Li H-X, Han L-H, Jiang W-P (2007) Mesenchymal stem cells transfected with HCN2 genes by LentiV can be modified to be cardiac pacemaker cells. Med Hypotheses 69:1093–1097

218. Tong M, Yang X-J, Geng B-y, Han L-H, Zhou Y-F, Zhao X et al (2010) Overexpression of connexin 45 in rat mesenchymal stem cells improves the function as cardiac biological pacemakers. Chin Med J 123:1571–1576

219. Chen L, Deng Z-J, Zhou J-S, Ji R-J, Zhang X, Zhang C-S et al (2017) Tbx18-dependent differentiation of brown adipose tissue-derived stem cells toward cardiac pacemaker cells. Mol Cell Biochem. doi:10.1007/s11010-017-3016-y

220. Wobus AM, Wallukat G, Hescheler J (1991) Pluripotent mouse embryonic stem cells are able to differentiate into cardiomyocytes expressing chronotropic responses to adrenergic and cholinergic agents and Ca2+ channel blockers. Differentiation 48:173–182

221. David R, Stieber J, Fischer E, Brunner S, Brenner C, Pfeifer S et al (2009) Forward programming of pluripotent stem cells towards distinct cardiovascular cell types. Cardiovasc Res 84:263–272

222. Burridge PW, Matsa E, Shukla P, Lin ZC, Churko JM, Ebert AD et al (2014) Chemically defined generation of human cardiomyocytes. Nat Methods 11:855–860

223. Hazeltine LB, Badur MG, Lian X, Das A, Han W, Palecek SP (2014) Temporal impact of substrate mechanics on differentiation of human embryonic stem cells to cardiomyocytes. Acta Biomater 10:604–612

224. Lian X, Hsiao C, Wilson G, Zhu K, Hazeltine LB, Azarin SM et al (2012) Robust cardiomyocyte differentiation from human pluripotent stem cells via temporal modulation of canonical Wnt signaling. Proc Natl Acad Sci U S A 109:E1848–E1857

225. Ojala M, Rajala K, Pekkanen-Mattila M, Miettinen M, Huhtala H, Aalto-Setala K (2012) Culture conditions affect cardiac differentiation potential of human pluripotent stem cells. PLoS One 7:e48659

226. Kleger A, Seufferlein T, Malan D, Tischendorf M, Storch A, Wolheim A et al (2010) Modulation of calcium-activated potassium channels induces cardiogenesis of pluripotent stem cells and enrichment of pacemaker-like cells. Circulation 122:1823–1836

227. Jara-Avaca M, Kempf H, Ruckert M, Robles-Diaz D, Franke A, de La Roche J et al (2017) EBIO does not induce cardiomyogenesis in human pluripotent stem cells but modulates cardiac subtype enrichment by lineage-selective survival. Stem Cell Rep 8:305–317

228. Scavone A, Capilupo D, Mazzocchi N, Crespi A, Zoia S, Campostrini G et al (2013) Embryonic stem cell-derived CD166+ precursors develop into fully functional sinoatrial-like cells. Circ Res 113:389–398

229. Rust W, Balakrishnan T, Zweigerdt R (2009) Cardiomyocyte enrichment from human embryonic stem cell cultures by selection of ALCAM surface expression. Regen Med 4:225–237
230. Hashem SI, Claycomb WC (2013) Genetic isolation of stem cell-derived pacemaker-nodal cardiac myocytes. Mol Cell Biochem 383:161–171
231. Rimmbach C, Jung JJ, David R (2015) Generation of murine cardiac pacemaker cell aggregates based on ES-cell-programming in combination with Myh6-promoter-selection. J Vis Exp 96:e52465
232. Wolfien M, Rimmbach C, Schmitz U, Jung JJ, Krebs S, Steinhoff G et al (2016) TRAPLINE: a standardized and automated pipeline for RNA sequencing data analysis, evaluation and annotation. BMC Bioinform 17:21

Adv Biochem Eng Biotechnol (2018) 163: 117–146
DOI: 10.1007/10_2017_24
© Springer International Publishing AG 2017
Published online: 8 December 2017

Bioengineered Cardiac Tissue Based on Human Stem Cells for Clinical Application

Monica Jara Avaca and Ina Gruh

Abstract Engineered cardiac tissue might enable novel therapeutic strategies for the human heart in a number of acquired and congenital diseases. With recent advances in stem cell technologies, namely the availability of pluripotent stem cells, the generation of potentially autologous tissue grafts has become a realistic option. Nevertheless, a number of limitations still have to be addressed before clinical application of engineered cardiac tissue based on human stem cells can be realized. We summarize current progress and pending challenges regarding the optimal cell source, cardiomyogenic lineage specification, purification, safety of genetic cell engineering, and genomic stability. Cardiac cells should be combined with clinical grade scaffold materials for generation of functional myocardial tissue in vitro. Scale-up to clinically relevant dimensions is mandatory, and tissue vascularization is most probably required both for preclinical in vivo testing in suitable large animal models and for clinical application.

M. Jara Avaca and I. Gruh (✉)
Leibniz Research Laboratories for Biotechnology and Artificial Organs (LEBAO), Department for Cardiothoracic, Vascular and Transplantation Surgery (HTTG), Hannover Medical School (MHH) & Cluster of Excellence REBIRTH, Hannover, Germany
e-mail: Gruh.Ina@mh-hannover.de

Graphical Abstract

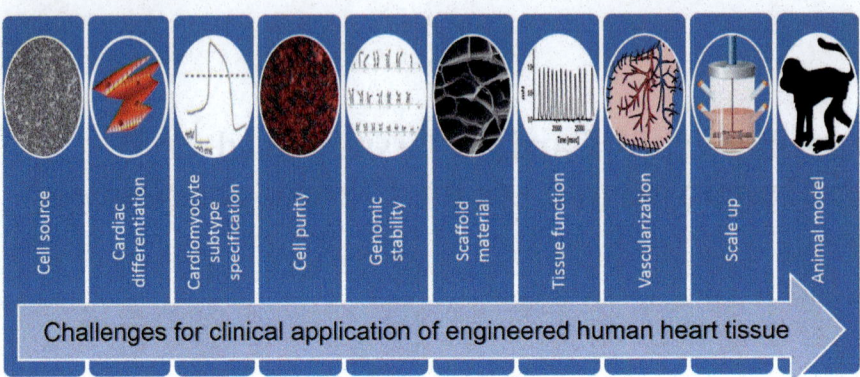

Keywords Cardiomyocytes, Myocardial tissue, Pluripotent stem cells, Scaffolds, Tissue engineering

Contents

Abbreviations

AAVS1	Adeno-associated virus integration site (safe harbor site)
ABCG2	ATP-binding cassette transporter protein
AMI	Acute myocardial infarction
ARVCM	Arrythmogenic right ventricular cardiomyopathy
ASC	Adipose tissue-derived cell
AV-block	Atrioventricular block
AV-node	Atrioventricular node
BCRP	Breast cancer resistance protein
BMP	Bone morphogenic protein

CD117	c-kit
CD106/VCAM-1	Vascular cell adhesion molecule 1
CD166/ALCAM	Activated leukocyte cell adhesion molecule
CD172A/SIRP-alpha	Signal regulatory protein alpha
CDC	Cardiosphere-derived cell
CMPM	Cardiac myocyte-populated matrix
c-Myc	Avian myelocytomatosis viral oncogene homolog
COUP-TF I and II	Chicken ovalbumin upstream promoter transcription factor I and II
CRISPR/Cas9	Clustered regularly interspaced short palindromic repeats/CRISPR-associated system
CRPC	Cardiac resident progenitor cell
CTLA4	Cytotoxic T-lymphocyte-associated protein 4
DNA	Deoxyribonucleic acid
EBIO	1-Ethyl-2-benzimidazolinone
ECM	Extracellular matrix
eGFP	Enhanced green fluorescent protein
EHT	Engineered heart tissue
ESC	Embryonic stem cell
FACS	Fluorescence-activated cell sorting
FGF-16	Fibroblast growth factor 16
GATA4	GATA-binding protein 4
GFP	Green fluorescent protein
hESC	Human embryonic stem cell
hiPSC	Human induced pluripotent stem cell
HUVEC	Human umbilical vein endothelial cell
ICF	Immunodeficiency, centromeric region instability, facial anomalies
iPSC	Induced pluripotent stem cell
IWP	Inhibitor of Wnt production
Klf4	Kruppel-like factor 4
$lin^{neg}/c\text{-}kit^{pos}$	CD31, CD34, CD45 negative/CD117 positive
LVEF	left ventricular ejection fraction
Meis-1	Meis homeobox 1
miR-128	micro RNA 128
MLC2a	Myosin light chain 2a
MLC2v	Myosin light chain 2v
MRI	Magnetic resonance imaging
MSC	Mesenchymal stem cell
MYDGF	Myeloid-derived growth factor
Nkx2.5	NK2 homeobox 5
NRCM	Neonatal rat cardiomyocytes
NRG1β/ERBB	Neuregulin 1/estrogen receptor beta

Oct4	Octamer-binding protein 4
p38 MAPK	p38 mitogen-activated protein kinase
PCR	Polymerase chain reaction
PDGFRβ	Platelet-derived growth factor receptor β
PSC	Pluripotent stem cell
RGD	Arginyl-glycyl-aspartic acid motif
sca-1	Stem cell antigen-1
Sox2	Sex determining region Y-box 2
SP	Side population
TALEN	TAL effector nuclease
TnI	Troponin I
USSC	Unrestricted somatic stem cell
VSD	Ventricular septal defect
Wnt	Wingless protein
ZFN	Zinc-finger nuclease
α-MHC, MYH6	α-myosin heavy chain promoter

1 Introduction

The clinical need for stem cell-based engineered cardiac tissue is increasing with the number of patients suffering from cardiovascular diseases. This trend is no longer restricted to industrialized countries, but is also becoming more and more evident in developing countries [1]. Potential applications for cardiac tissue range from cardiac malformations in newborns via hereditary cardiomyopathies to acute life-threatening conditions after myocardial infarction. During myocardial infarction, for example, a coronary artery occlusion leads to acute hypoxia in the myocardium with rapid loss of viable cells. Endogenous myocardial regeneration is limited [2, 3], mainly because of the limited proliferation potential of postnatal cardiomyocytes in mammals, including humans [4].

Instead, the dying cardiomyocytes are replaced by noncontractile fibrotic scar tissue and the remaining myocardium has to compensate for the loss of contractile function, which can lead to dilation (typically of the left ventricle) and subsequent heart failure [5]. The ultimate treatment option is heart transplantation. Although this leads to complete recovery of contractile function, it is associated with the problems typical of organ transplantation, such as donor organ shortage and the need for life-long immunosuppression.

In vitro engineered heart tissue (EHT) could provide an alternative treatment option for a number of cardiovascular diseases [6–8]. Because of the structural and functional complexity of the heart, the generation of transplantable organs as a whole is not a realistic goal in the near future. Nevertheless, bioartificial myocardial patches might be suitable for replacing damaged or diseased heart tissue and restoring contractile function. The features of in vitro engineered myocardial tissue necessary to achieve this are detailed below and summarized in Table 1. An obvious

Table 1 Desirable features of in vitro engineered myocardial tissue from human pluripotent stem cells for therapeutic purposes

Desired tissue property	Measureable	Biological or technical prerequisite
Sufficient active force	– Contraction force [mN] – Contraction frequency [Hz] – Contraction velocity	– Expression of proteins of the contractile apparatus – Alignment of cells in the tissue
Sufficient passive force	– Passive force [mN]	– Stable extracellular matrix (added or produced by cells)
Physiological function	– Positive force-preload relation – Positive force-frequency relation – Reaction to adrenergic and cholinergic stimulation – Positive inotropic effect of increased calcium concentration	– Alignment of actin and myosin filaments – Expression of adrenergic receptor – Expression of muscarinergic receptors – Functional sarcoplasmic reticulum
Electromechanical coupling	– Synchronous calcium oscillations – Conduction velocity [cm/s]	– Expression of ion channels – Expression of connexins
Connectivity to the host vasculature	– Survival of tissue after transplantation – Histological assessment of vascularization	– Endothelial cells – Smooth muscle cells or – Angiogenic factors
No rejection	– Survival of tissue after transplantation – Histological assessment	– Human/autologous cells – Non-immunogenic, biocompatible materials and media: non-xenogenic materials
No tumor formation	– Purity of cardiomyocytes – Assessment of tumor formation in animal models	– Efficient exclusion of undifferentiated cells – Unmodified or safely modified and genetically stable cells
No arrhythmogenic potential	– Purity of ventricular cardiomyocytes – Assessment of arrhythmias in animal models	– Efficient exclusion of pacemaker-like cells – Efficient maturation and integration of ventricular-like cells

Properties related to safety of clinical application are highlighted in bold

advantage of tissue engineered patches compared with cellular therapies is the possibility of treating larger defects, potentially including congenital malformations. Ideally, the use of nonimmunogenic cell sources and materials should obviate the need for immunosuppression.

1.1 Clinical Need for Engineered Cardiac Tissue

Cardiovascular diseases account for more than 4 million deaths in Europe each year; they are the leading cause of morbidity and mortality and have a massive socioeconomic impact [9]. Tissue damage in the left ventricle as a result of acute myocardial infarction (AMI) is the most common of many clinical conditions calling for new therapeutic approaches based on engineered cardiac tissue. Arrythmogenic right ventricular cardiomyopathy (ARVCM) is characterized by loss of healthy cardiomyocytes and replacement of contractile myocardium with fatty tissue. Similar to the fibrous scar following AMI, this fat is noncontractile and impairs normal conduction in the heart. The resulting arrhythmias can be life threatening, and ARVCM progression can lead to heart failure [10]. In vitro-generated myocardial tissue could be used for surgical replacement of scar or fat tissue to restore the contractile function of the heart. It might also be a treatment option after surgical excision of cancer tissue. However, it should be noted that the heart is rarely affected by primary or secondary tumors. Primary cardiac tumors account for only 5–10% of all tumors in the heart [11], whereas metastases originating from other primary tumors are much more common [12]. The most common primary cardiac tumors are atrial cardiac myxomas, which are mostly benign and often found incidentally [13], with a reported prevalence of 0.0017–0.19% at autopsy [14]. About 25% of primary cardiac tumors are malignant and, of these, 75% are sarcomas [12] requiring surgical treatment, which could be combined with transplantation of bioengineered tissue. In contrast, for primary cardiac lymphomas, chemotherapy and/or irradiation are preferred treatment options [15].

Congenital defects of the heart are frequently related to malformation of vessels and valves, but may also include the lack of contractile myocardial tissue. The most common defects are ventricular septal defects (VSD), which account for up to 40% of all congenital heart malformations [16]. Although the septal wall is mainly composed of muscular tissue, surgical closure of the defect with a noncontractile patch, using materials such as (autologous) pericardium or Dacron, is considered safe and very effective, with 99.5% of patients being asymptomatic 2 years after surgery [17]. However, implantation of in vitro-generated myocardial tissue might be a treatment option for malformations affecting the ventricles themselves. In patients with hypoplastic left heart syndrome, an underdeveloped left ventricle is not capable of supporting sufficient circulation because of its reduced size [18]. This condition can develop as a consequence of valvular problems [19], but occurs in a subset of patients as isolated hypoplastic left heart complex without malformation of the valves [20], which could potentially benefit from transplantation of EHT.

1.2 Myocardial Tissue Engineering Strategies

Engineered heart tissue can be created in vitro from cardiomyocytes together with other cell types by combining them with different types of scaffolds of various

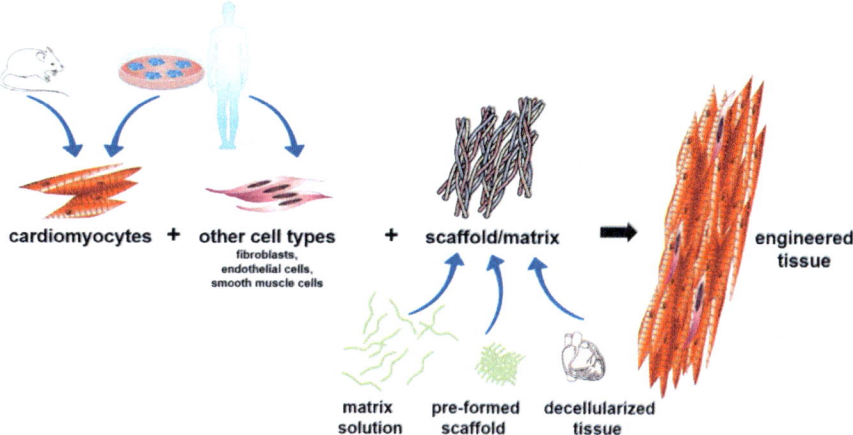

cardiomyocytes + other cell types + scaffold/matrix ➡ engineered
           fibroblasts,   tissue
           endothelial cells,
          smooth muscle cells

matrix   pre-formed   decellularized
solution   scaffold   tissue

Fig. 1 Principle of myocardial tissue engineering. Cardiomyocytes (either animal-derived primary cells or cardiomyocytes differentiated from hPSCs) can be combined with other cell types and different types of scaffolds (to provide an extracellular matrix) for the in vitro generation of engineered tissue. Experimental strategies include mixing of cells with aqueous solutions of matrix components for de novo tissue self-assembly in casting molds, and cell seeding onto preformed natural or synthetic scaffolds (including decellularized tissue)

compositions and origins. The cells can be either mixed with solutions of matrix components for in situ tissue formation in casting molds or seeded onto preformed natural or synthetic scaffolds, including decellularized tissue (Fig. 1). Before pluripotent stem cell-derived cardiomyocytes became a viable therapeutic option, experimental strategies for the generation of engineered myocardial tissue were established using primary heart cells from different species. In 1997, Eschenhagen et al. demonstrated the use of embryonic chicken cardiomyocytes together with a collagen matrix, forming beating cardiac myocyte-populated matrices (CMPMs) [21]. This was the basis for the creation of beating three-dimensional EHT from neonatal rat cardiomyocytes (NRCMs) as an in vitro model for drug response [6] or for diseases such as stress-related hypertrophy [22]. These studies featured in situ tissue formation after a casting process of mixing cardiomyocytes from dissociated heart tissue with a solution of extracellular matrix (ECM) components, in contrast to NRCM seeding onto preformed collagen matrices [23]. Rat EHT shows similar morphological features to native heart tissue, with aligned cardiomyocytes showing features such as organized sarcomere structures, gap junctions, and T-tubuli. The cardiomyocytes have spontaneous beating activity with recordable action potentials and measurable twitch (i.e., contraction) forces responding to beta-adrenergic inotropic stimulation [24]. The contractile forces of engineered tissues have been measured directly after transfer to an organ bath and connection to force transducers [6] or during cultivation in a bioreactor using special culture vessels connected to a force sensor [25]. Indirect measurement can be made using video-optical analysis of beating tissue such as fibrin-based mini-EHTs, which have been proposed as a high-throughput drug screening platform [26]. Scale-up of EHT has also been achieved by fusion of individual EHTs to bigger constructs using primary cells from rat heart [27].

1.3   Cell Sources for Myocardial Repair

The final aim of myocardial tissue engineering is the generation of contractile tissue constructs for treating patients; therefore, only human cardiomyocytes are a suitable cell source. Ideally, they should be patient-derived; however, primary human cardiomyocytes cannot be used because of their limited proliferation capacity [4], preventing in vitro expansion to cell numbers required for engineering constructs of clinically relevant dimensions. Alternative cell sources for cardiac regeneration and myocardial tissue engineering are presented in the following subsections.

1.3.1   Adult Stem Cells

Adult stem cells have the advantage that they can be derived from the patient and therefore represent a potentially autologous cell source for regenerative therapies without the need for immunosuppression. However, there is an ongoing debate about whether they hold the potential to differentiate into bona fide cardiomyocytes.

Early studies reported beneficial effects following intracoronary transplantation of autologous bone marrow-derived cells [28]. Studies in mice claimed that the regenerative effect of bone marrow cells locally delivered to injured hearts was a result of de novo generation of contractile myocardium by "transdifferentiation" of CD31, CD34, CD45 negative/CD117 positive (lin^{neg}/c-kit^{pos}) bone marrow progenies [29]. This notion was challenged by later studies showing no evidence for the generation of cardiomyocytes in the infarcted region of the myocardium in a similar setting. These studies suggest that rare events of cell fusion could account for the presence of cardiomyocytes seemingly derived from bone marrow [30]. Moreover, a positive effect of bone marrow cells on heart regeneration in humans, suggested in an early study with only a few patients [31], could not be confirmed in a long-term follow-up of a larger randomized and controlled clinical trial. Although early improvement of diastolic function was observed 6 months after transplantation in the randomized trial, this effect did not lead to a sustained effect after 60 months [32, 33]. It was hypothesized that the transplanted stem cells release soluble factors that act in a paracrine fashion to support cardiac regeneration [34]. Secretome analyses of bone marrow-derived cells identified paracrine-acting proteins, such as myeloid-derived growth factor (MYDGF), which promoted cardiac myocyte survival and angiogenesis in a mouse model [35]. Moreover, in experiments aiming at the in vitro differentiation of mesenchymal stroma cells (MSCs) into cardiomyocytes, some groups showed cardiac marker expression in derivatives, including [36] or excluding the induction of action potentials (spontaneous or stimulated) [37]. It is therefore still controversial whether stromal cells originating from different tissues have the potential to differentiate into bona fide cardiomyocytes (also reviewed in [38]).

Other studies reported cardiac differentiation of stem cell populations resident in the adult myocardium [39]. Different types of these cardiac resident progenitor cells (CRPCs) were identified on the basis of expression of the stem cell factor receptor c-kit (CD117) in rat [40] and human myocardium [41–44] or expression of stem cell antigen-1 (sca-1) [45, 46]. Although these cell types are referred to as "stem cells," they do not fulfil all necessary criteria, such as unlimited self-renewal, clonal expansion, and multipotentiality [47]. On the one hand, autologous cardiac c-kit^{pos} cells have been used in clinical trials in patients with heart failure after myocardial infarction and resulted in improved LV systolic function and reduced infarct size [48]. On the other hand, the cardiomyogenic nature of endogenous c-kit^{pos} cardiac cells is still questioned by others [49]. A recent review suggests that the apparently discrepant results of individual studies might be because there is no homogeneous c-kit^{pos} cell population but distinct populations with different origins and different plasticities. These populations include heart field progenitors on the one hand and epicardium-derived, noncardiomyogenic precursors with a mesenchymal phenotype on the other hand [50].

The heart also contains cells with a "side population" (SP) phenotype [51], which is defined by the capability to exclude Hoechst dye through the expression of ATP-binding cassette transporter protein (ABCG2) [52] (also known as breast cancer resistance protein; BCRP) [53, 54]). The potential to differentiate into beating cardiomyocytes in vitro was reported for rat SP cells [55], but could not be confirmed for a human cell population [44]. Another progenitor cell type derived from the heart, termed "cardiospheres" or "cardiosphere-derived cells" (CDCs) has been expanded in vitro from myocardial biopsy material [56]. Regenerative potential after transplantation was also reported for this cell type in different animal models [57, 58] and a first clinical trial demonstrated that intracoronary administration of autologous CDCs is safe. Moreover, therapeutic effects on scar size and regional myocardial function at 1 year post-treatment were reported [59]. Nevertheless, a mainly indirect mechanism of action has also been proposed for CDCs, whereby cardiac regeneration may be the result of secreted factors. However, some degree of direct contribution to the formation of cardiomyocytes was claimed by the investigators [60].

Literature on the use of adult cardiac stem cells for tissue engineering is scarce. Some studies have claimed cardiac differentiation of murine CDCs (i.e., upregulation of cardiac marker expression) after combining them with scaffolds such as electrospun polymers with defined mechanical properties [61] or after encapsulation in hydrogels and cultivation under "native heart-mimicking dynamic stretch environment" conditions [62]. Similar observation of early cardiac marker expression was reported for human CDCs cultivated on an anisotropically nanopatterned surface (providing a defined orientation to the attached cells), which was therefore proposed for therapeutic application as a patch [63]. Similarly, human cardiospheres showed upregulated expression of GATA4, Nkx2.5, and TnI cardiospheres in the presence of collagen-based RGD-functionalized scaffolds [64].

In summary, a number of adult stem or progenitor cell types have been investigated for myocardial repair with promising results. Evaluation of their potential for differentiation into cardiomyocytes in vitro and in vivo is ongoing. Importantly, to date there are no reports demonstrating the generation of adult stem cell-derived cardiomyocytes of sufficient quality and quantity to enable generation of contractile human cardiac tissue.

1.3.2 Pluripotent Stem Cells

In contrast to different types of adult stem cells, pluripotent stem cells (PSCs), including embryonic stem cells (ESCs) [65] and induced pluripotent stem cells (iPSCs) [66, 67], show almost unlimited potential for self-renewal and differentiation. Therefore, they are attractive cell sources for the treatment of tissue defects. iPSCs can be derived from various sources and cell types from the adult organism, including blood cells and keratinocytes [66, 68, 69], and offer the unique opportunity of using patient-specific cells for individualized treatment. Most importantly, iPSCs can be differentiated into functional cardiomyocytes, which was demonstrated both for mouse iPSCs [70–72] and their human counterparts [68] and unequivocally corroborated in a number of studies.

However, iPSCs generated by overexpression of exogenous factors using gamma-retroviral vectors (the original strategy used for somatic cell reprograming) cannot be implemented for therapeutic applications. This issue was successfully addressed by the reduction of factors needed for reprogramming somatic cells [73–76], assisted by small molecules compensating for the reduced efficiency that usually occurs after omission of one or two factors [77]. Moreover, drug-inducible systems for the expression of transgenes leading to reprogramming have been described [78, 79] as well as induction of iPSCs using nonintegrating viral vectors, such as adenoviruses [80] or sendai viruses [81, 82] transiently expressing Oct4, Sox2, Klf4, and c-Myc. Nonviral plasmids have also been used to induce virus-free iPSCs [83]. These approaches have also been used in combination with systems enabling excision of the reprogramming factors once reprogramming has occurred [84] or with temperature-sensitive Sendai variants for rapid elimination of remaining viral vector-related genes [85]. Thus, these transgene-free human iPSCs (hiPSCs) provide a unique cell source for the generation of bioengineered cardiac tissue for clinical application.

2 Challenges for Clinical Translation of Stem Cell-Derived Cardiac Tissue

Before clinical application of engineered cardiac tissue based on human PSCs can be realized, a number of limitations still have to be addressed (summarized in the graphical abstract). We outline current progress regarding the safety of genetic cell engineering, cardiomyogenic lineage and subtype specification, and cardiomyocyte purification. For in vitro generation of functional myocardial tissue, cardiac cells should be combined with clinical-grade scaffold materials. In addition, up-scaling to clinically relevant dimensions is mandatory, probably requiring tissue vascularization both for preclinical in vivo testing in suitable large animal models and for clinical application.

2.1 Genome Engineering and Integrity of Pluripotent Stem Cells

For potential clinical application of iPSCs, controlled and safe transgene integration or footprintless gene editing are desirable from a safety perspective, in addition to transgene-free approaches for iPSC generation. When prolonged expression of transgenes is required, for example to enable efficient cell type selection (discussed in detail in Sect. 2.3), targeted integration into "safe harbor sites" such as the AAVS1 locus is a promising approach for controlled and safe insertion. This can be achieved using designer nucleases such as the zinc-finger nucleases (ZFNs) [86], TAL effector nucleases (TALENs) [87], or the CRISPR/Cas9 system [88]. For details on current techniques for site-specific genome engineering in PSCs, please see Merkert et al. [89] and Merkert and Martin [90] (in this issue). In addition, current research is focused on the (epi-)genomic stability of reprogrammed cells, as chromosomal aberrations have been reported and might pose the risk of cell transformation and subsequent tumor formation [91]. The extent of risk associated with clinical translation of engineered cardiac tissue must be investigated carefully for mutations that (1) result from pre-existing mutations in somatic cells even before reprogramming, (2) are introduced as de novo mutations during cell reprogramming, and/or (3) accumulate during the extensive in vitro proliferation of hPSCs required for therapeutic applications [92].

2.2 Lineage and Cardiomyocyte Subtype Specification

Early protocols for the induction of cardiac differentiation of hPSCs relied on undefined culture conditions and/or supplements. Examples include cell cultivation in medium supplemented with pretested batches of fetal calf serum [93], on supportive stromal layers [94], or with conditioned medium [95], either alone or in combination with molecular inhibitors of p38 MAPK [96], resulting in heterogeneous cell populations requiring further cardiomyocyte purification (e.g., by antibiotic selection) [97]. Recent years have brought additional insight into the individual steps underlying the pathways and molecular modulators of lineage specification: from the PSC state toward mesoderm progenitor cells, cardiac mesoderm, and cardiac progenitor cells into early cardiomyocytes and their further maturation [98]. Consequently, specific modulation of the canonical Wnt signaling pathway with small molecule inhibitors such as CHIR99021 (a glycogen synthase kinase 3 inhibitor) and IWP (inhibitor of Wnt production) has been used to obtain robust cardiomyocyte differentiation for many different human ESC and iPSC lines under fully defined conditions [99, 100].

The first reports of cardiac differentiation of hiPSCs readily demonstrated the presence of ventricle-like and pacemaker-like cells with distinct action potential characteristics, as shown by patch clamp analyses [68]. Pacemaker-like cells (often referred to as nodal-like cells) were identified in differentiating hESCs by their GATA6-GFP expression and could be enhanced in numbers by inhibition of NRG1β/ERBB signaling [101]. Interestingly, Ben-Ari et al. reported that iPSC-cardiomyocyte populations shifted in the course of development from a nodal-like to an atrial/ventricular-like phenotype, and also described transitional populations [102]. Investigating the proportions of cardiomyocyte subtypes, Lian et al. reported up to 98% cardiomyocytes, with more than 50% of them expressing MLC2v [99]. However, in the early phases of differentiation, MLC2v^{pos} cardiomyocytes frequently co-expressed MLC2a, therefore no clear segregation of atrial and ventricular markers could be observed [103]. Burridge et al. reported optimized cardiac differentiation efficiencies of 80–95% for different ESC and iPSC lines, with a progressive decrease in MLC2a^{pos} atrial-like cells in favor of higher numbers of MLC2v^{pos} ventricular-like cells (up to 60% at day 60). This was confirmed by patch clamp analysis, showing 57% ventricular-like cells together with atrial-like and nodal-like cells [104]. Kempf et al. showed that a protocol using targeted modulation of the Wnt pathway with small molecules in suspension cultures typically resulted in 80–90% ventricular cardiomyocytes [105] (see also Kempf and Zweigerdt in this issue [106]).

Despite the longstanding knowledge about the presence of different cardiomyocyte subtypes following differentiation, only recently have novel approaches for selective differentiation emerged. Application of 1-ethyl-2-benzimidazolinone (EBIO), a chemical modulator of small/intermediate-conductance Ca^{2+}activated potassium channels (SKs 1–4), at early time points of cardiac differentiation induced cardiomyogenesis and enriched nodal-like cells of murine PSCs [107]. However, in the human system, only lineage-selective survival of cardiac progenitors was observed after EBIO application, finally producing higher proportions of cardiomyocytes with shorter AP durations [108]. Furthermore, large amounts of atrial-like cells were induced from hESCs by supplementation of retinoic acid [109], whereby pathway inhibition of the retinoic acid receptor antagonist BMS-189453 during differentiation led to enhanced amounts of the ventricular-like cardiomyocyte subtype. Along the same line, orphan nuclear receptor transcription factors COUP-TF I and II were found to regulate atrial identity. Induction with retinoic acid led to 85% atrial-like hESC-derived cardiomyocytes, which were proposed for use as a preclinical model for atrial-selective pharmacology [110]. Similarly, Protze et al. used very precise control of bone morphogenic protein (BMP) and Wnt signaling with defined concentrations of

BMP, Activin A, and retinoic acid (among others) to generate human sinoatrial node-like progenitor cells that were able to rescue an experimentally induced AV block after transplantation in mice [111].

These novel approaches offer a promising avenue for tailored regenerative strategies targeting ventricular and atrial myocardial defects and even impaired myocardial conduction, which could be treated by the generation of PSC-derived nodal/pacemaker cells or tissue constructs for the reconstruction of AV nodes.

2.3   Cell Purification for Safe Transplantation

Purification of hPSC-derived cardiomyocyte populations is mandatory for generating functional myocardial replacement tissue without any contaminating extracardiac components and with a limited risk of teratoma formation following transplantation. Early studies used murine ESC-derived cardiomyocytes purified (or rather enriched) by density gradient centrifugation using a Percoll solution for the generation of contractile tissue [112, 113]. Only recently have surface markers expressed on cardiomyocytes been described and implemented for cardiomyocyte purification (i.e., CD172A/SIRP-alpha [114], and CD166/VCAM-1 [115]). Moreover, selective culture conditions for cardiomyocyte enrichment by medium supplementation with lactate instead of glucose [116–118] and fluorescence-activated cell sorting (FACS) after selective labeling with mitochondria-specific dyes have been demonstrated [119].

Cardiomyocytes can be purified efficiently from transgenic cell lines expressing antibiotic resistance genes under the transcriptional control of cell type-specific promoters. For cardiomyocytes, the α-myosin heavy chain promoter (α-MHC, MYH6) has found widespread use [120–122]. Purified cardiomyocytes selected by the addition of antibiotics following differentiation of genetically engineered hPSCs were used together with collagen or fibrin-based matrices for tissue generation [123–125]. Our own work demonstrated that such antibiotic selection of transgenic cardiomyocytes could be performed with high efficiency directly on cell aggregates after differentiation in embryoid bodies. The resulting "cardiac bodies" displaying cardiomyocyte purities of >99% can be directly used to generate bioartificial cardiac tissue without the need for single cell dissociation [126]. We have now combined this purification strategy with TALEN-mediated targeted transgene integration into the AAVS1 locus of a selection cassette (α-MHC promoter-driven neomycin resistance) to generate stable hiPSC lines. These lines allow efficient cardiomyocyte selection by medium supplementation with G418 for the production of bioartificial cardiac tissue with precise control of cellular composition (Fig. 2).

Novel approaches for cardiomyocyte selection such as miRNA switches (i.e., synthetic RNAs that can "sense" cardiomyocyte-specific microRNAs, thereby enabling cardiomyocyte purification) [127], have not been used for tissue engineering applications. To exclude potential teratoma formation from residual pluripotent

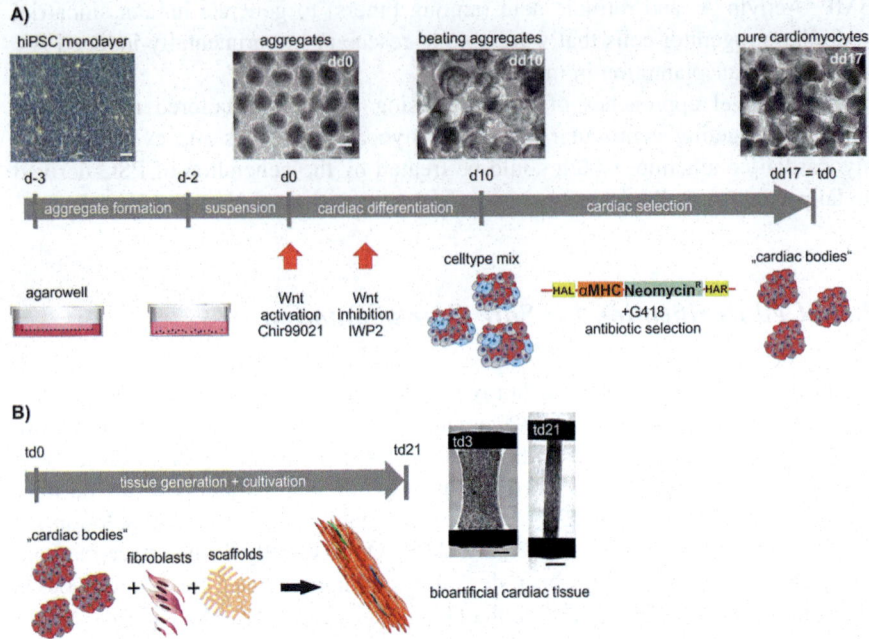

Fig. 2 Overview of hiPSC cardiac differentiation and tissue production. (**a**) For aggregate formation, hiPSC monolayer (*scale bar* 200 μm)-derived single cells are inoculated into agarowells. Differentiation day 0 (*dd0*) aggregates (*scale bar* 200 μm) are transferred after 24 h to suspension culture and differentiated under agitated conditions. Application of Wnt pathway activator Chir99021 followed by Wnt pathway inhibitor IWP2 results in beating aggregates on d10 (*scale bar* 200 μm), composed of cardiomyocytes and other cell types. After application of the antibiotic G418, dd17 aggregates (*scale bar* 200 μm) consist only of pure cardiomyocytes and are termed "cardiac bodies." The selection strategy is based on AAVS1 locus targeted integration of a tissue-specific promoter (αMHC)-driven neomycin resistance. (**b**) Mixing cardiac bodies with fibroblasts and extracellular matrix (scaffolds) into a specific casting mold allows in vitro cardiac tissue production. Bioartificial cardiac tissue samples (*scale bars* 1,000 μm) are cultivated for up to tissue day (*td*) 21 followed by mechanical, physiological, and structural characterization

cells [122], an innovative negative selection strategy has been recently proposed. It relies on iPSC-eliminating agents and involves overexpression of alkaline phosphatases on iPSCs (termed "ecto-alkaline phosphatase"), which induce cell death selectively in undifferentiated iPSCs but not in iPSC-derived cardiomyocytes upon the addition of synthetic peptides [128].

2.4 Clinical Grade Scaffold Materials

Human pluripotent stem cell-derived cardiomyocytes of different purities have been used for cardiac tissue engineering together with a variety of scaffold materials, such as open porous polymer matrices [129] [130], photopolymerizable hydrogels based on polyethyleneglycol and fibrinogen [131], and poly(glycerol-sebacate) [132], or without matrix [133, 134]. Such thin sheets of matrix-free tissue have been produced from selected cardiomyocytes using the "cell sheet" technology [135]. In this approach, complete cell layers can be detached from a thermoresponsive cell culture surface by a simple change of the cultivation temperature [136]. The technology led to the formation of contractile cardiac tissue in vitro, but direct comparison of tissue functionality between individual studies is difficult because in many cases contraction forces were not measured. Contraction forces were first reported by Schaaf et al. for hESC-derived EHT using Matrigel™ (derived from murine cells) and bovine fibrin [137]. Similarly, Tulloch et al. used murine Geltrex™ and rat collagen type I for the creation of functional tissue from ESC- and iPSC-derived cardiomyocytes. Our own work showed the formation of functional human bioartificial cardiac tissue not only using Matrigel™ and rat collagen type I, but also using a combination of chemically modified (crosslinkable) hyaluronic acid and human collagen type I, thus going another step forward in the direction of clinically applicable materials that are free of animal-derived components [126]. However, there is an ongoing debate about the requirement for clinical-grade scaffold materials and whether they have to be completely free of animal-derived components. Therefore, state-of-the-art protocols for the generation of human EHT still feature the use of Matrigel™ and bovine fibrinogen [138] or propose the use of medical grade bovine collagen without Matrigel™ [139].

Notably, not all approaches aim at using a scaffold merely as an ECM substitute to allow generation of native-like myocardial tissue for replacement therapy; for example, Chen et al. proposed the use of an elastomeric patch made from poly (glycerol-sebacate) for delivery of embryonic stem cells to the heart to provide initial mechanical support and subsequent release of cardiomyocytes to allow integration into the host myocardium [132].

2.5 Engineered Cardiac Tissue Function

Ideally, in vitro engineered myocardial tissue should closely resemble native myocardium in terms of mechanical function (i.e., it should be able to exert contractile force). In the clinical setting, cardiac performance is described mainly using hemodynamic parameters such as peripheral blood pressure [140]. Critical parameters in addition to beating frequency are the pumping capacity (determined as "left ventricular ejection fraction," LVEF), and data on pressure and volume relations in the heart (end-systolic volume, end-systolic pressure, end-diastolic volume, end-diastolic pressure) and their rate of change [141]. Detailed data concerning the underlying mechanics of cardiac contractions can be acquired using modern imaging techniques such as magnetic resonance imaging (MRI), which can visualize even local contractions and their impairment in a diseased heart (e.g., regional wall motility, wall thickness before and during contraction) [141].

For myocardial tissue engineering, "technical" parameters rather than "clinical" parameters are more relevant for measuring functionality; myocardial contractility and tissue mechanics are of major interest. During embryonic development, the contraction velocity of the heart increases over time [142]. However, for myocardial tissue, strain is not age dependent and has been determined as $32.7 \pm 10.7\%$ for the right ventricle and $23.1 \pm 9.1\%$ for the left ventricle [143]. The underlying contraction forces were assessed in children using heart muscle samples obtained during surgical correction of congenital malformations. Wiegerinck et al. demonstrated that a newborn's right ventricle (at the age of <2 weeks) showed contraction forces of up to 1.4 ± 0.3 mN/mm^2 whereas infants (3–14 months) displayed higher values of up to 1.7 ± 0.9 mN/mm^2 [144]. Interestingly, data from this study of 3-month-old infants also showed an increased contraction force at higher beating frequencies (i.e., a positive force–frequency relation), which is not present in newborns and can be considered one hallmark of postnatal tissue maturation. The active contraction force of isolated heart muscle strips has been reported to range between 14.5 ± 4.4 mN/mm^2 and 22.8 ± 1.4 mN/mm^2 for adult humans [145, 146]. For patients who underwent heart transplantation because of severe heart failure, contraction forces of ~4 mN/mm^2 have been determined in the explanted diseased myocardium [147].

In addition to active contraction forces, the passive forces of the tissue are also important because they define the stiffness of the material and thereby impact overall contractility. Passive forces in healthy human heart tissue were determined to be 11.3 ± 1.3 mN/mm^2 and, in contrast to active forces, are increased in chronic heart failure [148]. Another feature closely related to mechanical function is the anisotropy of the myocardium (i.e., the fact that most of its properties depend on orientation). This is true for the active contraction forces, which are maximized because of the parallel arrangement of cardiomyocytes and the resulting uniform uni-axial direction of contraction in the heart [149], as well as in myocardium generated in vitro [150]. At the same time, passive forces are directionally dependent, with higher values for measurement in the circumferential direction than in the longitudinal direction [151].

As stated above, contraction forces were not evaluated in most early studies on hESC-derived cardiac tissue. In later approaches, Schaaf et al. reported 0.12 mN/mm^2 for hESC-derived EHT [137] as compared with the results of Tulloch et al. (~0.08 mN/mm^2) [152], Kensah et al. (4.4 mN/mm^2), and Zhang et al. (11.8 ± 4.5 mN/mm^2) [153]. Recently, Jackman et al. combined hPSC-derived cardiomyocytes with a fibrin-based hydrogel to form very thin tissue strips, called "cardiobundles," with a diameter of ~0.03 mm^2. Under optimized culture conditions (dynamic culture for improved medium supply), these constructs showed a contraction force of 23.2 mN/mm^2 [154], closely resembling forces of the native human adult myocardium. For an overview on pharmacological force regulation and the effect of different culture conditions on several contractile parameters, please see Mannhardt et al. [155]. Using another approach for functional testing, Seta et al. produced a tubular construct with a wall thickness of 0.5 mm from cell sheets of hiPSC-derived cardiomyocytes and could demonstrate its function in vivo after transplantation around the inferior vena cava in nude rats, where it generated a pulse pressure of about 9 mm Hg upon electrical stimulation [156].

Similarly, considerable progress has been made toward improved function of engineered cardiac tissue in terms of electrophysiology. One exemplary parameter, the cardiac conduction velocity, was assessed in various studies using either microelectrode-array measurements or optical mapping of cell membrane potentials and/or calcium transients and subsequently optimized. In our own work, the conduction velocity in human bioartificial cardiac tissue was up to 4.9 cm/s [126], whereas Nunes, et al. demonstrated that electrical stimulation of three-dimensional, aligned cardiac tissue ("biowires") resulted in increased myofibril ultrastructural organization, ~40 and ~50% higher conduction velocity (with 3 and 6 Hz stimulation, respectively) of up to ~15 cm/s, and improved electrophysiological and Ca^{2+} handling properties [157]. Zhang et al. showed correlation between cardiomyocyte purity and conduction velocity, reaching 25.1 cm/s in patches of 90% cardiomyocytes [153], which is similar to the conduction velocity of 25.8 cm/s measured by Jackmann et al. [154]. Both rigorous cell purification to eliminate pacemaker-like cells and careful electrophysiological characterization of engineered cardiac tissue are needed to exclude an arrhythmogenic potential.

2.6   Upscaling Tissue Dimensions and Vascularization

In contrast to in vitro models of human stem cell-derived EHT calling for miniaturization to enhance throughput, future in vivo application of EHT for reconstructive therapy in a (pre-)clinical setting requires massive scale-up of tissue dimensions. Murry et al. estimated that one billion viable cells are lost after myocardial infarction [158]. Cardiac MRI revealed that, even after successful treatment of the acute infarction with primary percutaneous coronary intervention (PCI) or thrombolysis, the infarcted area accounts for $12.5 \pm 6.3\%$ to $22.6 \pm 12.3\%$ of the total left ventricle's mass of ~125 g in the late phase 3 months after myocardial infarction [159]. Therefore, considering a myocardial density of 1.082 ± 0.003 g/cm^3 [160] and the strategy of full transmural replacement to provide a normal mean left ventricular myocardial thickness (LVMT) of 5.3 ± 0.9 mm for women and 6.3 ± 1.1 mm for men [161], tissue patches of 5–8 cm in diameter might be needed. This requires scale-up of cell production, including novel methods for high-throughput generation of uniform aggregates from hPSCs [162] and cell expansion and differentiation in fully controlled bioreactors [163, 164]. Tiburcy et al. recently presented a "clinical-sized large patch" produced from 40×10^6 cells in an 8 mL volume, resulting in $35 \times 34 \times 0.5$ mm construct dimensions [139].

For in vitro generation of functional heart tissue – especially in larger dimensions and increased thickness – sufficient vascularization by endothelial cells and smooth muscle cells is a crucial aspect, in addition to the contractile force of cardiomyocytes [8]. Although an oxygen diffusion limit of 100–200 μm has been described for in vitro cultivation of engineered myocardial tissue, with decreasing cell viability in deeper areas of larger constructs [165], Shimizu et al. reported a maximum height of only 80 μm for a layered cardiac construct to achieve sufficient survival following transplantation [166]. To support cardiomyocyte function in human EHT, primary cells have been used, including human umbilical vein endothelial cells and MSCs [152]. In a proof-of concept study using murine ESCs, the generation of a tissue sheet reassembled with defined populations of ESC-derived cardiovascular cell types resulted in improved neovascularization and improved heart function after transplantation in a rat myocardial infarction model [167]. In another approach to creation of hPSC-derived cardiac muscle patches, multiphoton-excited 3D printing of a native-like ECM scaffold was combined with seeding of cardiomyocytes, smooth muscle cells, and endothelial cells [168]. Nakane et al. combined hiPSC-derived cardiomyocytes, $CD144^{pos}$ endothelial cells, and $PDGFR\beta^{pos}/CD140b^{pos}$ vascular mural cells into large-format engineered cardiac tissue and demonstrated the importance of tissue geometry. Although a total tissue area of about 170 mm^2 could be produced, cell survival was impaired in more compact tissues. A mesh-like structure with a bundle width of about 0.5 mm supported cellular survival and function in vitro and after transplantation [169].

Although small capillaries might form readily via self-assembly, even with prevascularized cardiac constructs it is still unclear whether coupling to (or ingrowth of) host-derived vessels will be fast and efficient enough to allow survival of the large implants needed for substantial regeneration. To address this, Zhang et al. created a biodegradable scaffold from a citric acid-based elastomer that allowed cell seeding on the one hand and surgical anastomosis on the other hand because of a preformed built-in vascular structure. Their "AngioChip" with inlet and outlet dimensions of 100 μm × 200 μm was successfully connected to the femoral vessels in adult rat hind limbs and remained stable and patent for at least 1 week [170]. Alternatively, prevascularization of tissue constructs can be performed in vivo, but may require repeated surgical intervention. Komae et al. created a multilayer graft of hiPSC-derived cardiomyocytes in nude rats; in this case, six-layer sheets generated in vitro were first transplanted onto fat tissue of the lower abdomen and later resected together with the femoral arteries and veins to make transplantable grafts with connectable vessels [171].

In addition to vascular cell types, fibroblasts are an important component of native heart tissue and their supportive function has been described for the generation and function of engineered tissue [27]. Fibroblasts [172, 173], endothelial cells [174, 175], and mural cells [169] can be differentiated from iPSCs. However, these cell types can also be isolated directly from the adult body, propagated in cell culture, and used in an autologous setting, which might be an easier and safer alternative for therapeutic application of engineered vascularized tissue.

2.7   Small and Large Animal Models for In Vivo Testing

Suitable animal models are very important for evaluating the therapeutic potential of EHT and assessing the potential risks associated with cellular therapies and tissue transplantation. Apart from the risk of teratoma formation, which can be excluded by rigorous elimination of undifferentiated stem cells, the risk of arrhythmias induced by the automaticity of immature ventricular cardiomyocytes or co-transplanted nodal-like cells has to be considered. For in vivo testing of in vitro-generated myocardial tissue from rat cardiomyocytes or murine ESCs or iPSCs, tissue samples were transplanted into rodents [112, 120, 176–178] to demonstrate tissue integration. In addition, human cardiomyocytes derived from ESCs were injected into the hearts of mice with severe combined immunodeficiency [179] and rats. The results showed midterm survival of some human cardiomyocytes for several weeks [95]. However, physiological differences, such as the obviously disparate beating frequencies of human hearts and mouse or rat hearts, preclude functional and physiological integration of human cardiomyocytes into mouse or rat hearts [180–182]. In addition, formation of fibrous tissue was observed around these cellular transplants and could potentially impair cardiac conduction and function after tissue transplantation. Interestingly, functional coupling of hPSC-derived cardiomyocytes to the host myocardium after transplantation was recently demonstrated in guinea pigs as a result of higher similarity in electrophysiology, including beating rate [183–185]. However, the possibilities for surgical intervention, catheter-based application, and assessment of functional parameters are also limited in this model due to the small size of the animal and heart, indicating the need for suitable large animal models for preclinical testing.

Our own data and others' studies on transplantation of human cells into pig hearts demonstrate the limitations of this xenogeneic animal model: human cardiomyocytes showed poor survival despite immunosuppression of the recipient animal. Almost all studies that were able to show survival of human cells used cell types associated with immunosuppressive or at least immune-modulatory properties, such as MSCs [186] or similar cell types derived from adipose tissue (adipose tissue-derived cells, ASCs) [187, 188] or umbilical cord (unrestricted somatic stem cells, USSCs) [189, 190]. However, for the latter, transplant survival could not be confirmed by others [191] and is therefore controversial [192]. Increased transplant survival of hiPSC derivatives was confirmed after co-transplantation of MSCs [193]. In this study, undifferentiated hiPSCs were injected transendocardially into the border zone of a 7-day-old myocardial infarction. Using high dose immunosuppression with cyclosporine A and prednisolone, long-term survival of hiPSC derivatives for 15 weeks was demonstrated for the first time. Interestingly, after transplantation of these undifferentiated hiPSCs into the porcine myocardium, Templin et al. detected only human endothelial cell progenies, but no cardiomyocytes [193]. Survival of endothelial cells was also reported by other groups after transplantation of vascular iPSC derivatives [194, 195], whereas others reported the low survival of human cardiomyocytes applied to pig hearts [196]. The reasons for this putative lineage-selective survival are currently unclear. However, these studies suggest that use of this xenogeneic large animal model is challenging. A more extensive review on the use of pig models to investigate transplantation of hPSCs is given by Roberts et al. [197].

For an allogenic transplantation setting in pig hearts, porcine stem cell-derived cardiomyocytes are not yet available. Extensive research on the generation of porcine iPSCs only yielded partial reprogramming into iPSC-like cells [198–201]. However, convincing pluripotency was not demonstrated because of their dependence on persistent expression of reprogramming factors, and differentiation into functional cardiomyocytes was not achieved. This limitation is also true for other farm animals such as sheep [202].

In principle, dogs can serve as animal models of cardiovascular disease [203], but – in contrast to humans and pigs – they have an extensive network of collateral vessels providing blood supply to the myocardium. Therefore, induction of myo-cardial infarction in a defined area is particularly difficult to control [181, 204]. Fur-thermore, the generation of induced PSCs has been reported in dogs [205], but no protocols for directed cardiac differentiation have been published to date.

Based on these considerations, nonhuman primates have recently gained atten-tion as a potential model, not only for human cardiac disease, but also for the investigation of cellular therapies in preclinical studies. Chong et al. were the first to demonstrate engraftment of hESC-derived cardiomyocytes in the infarcted heart of pigtail macaques (*Macaca nemestrina*) [206]. To allow survival of human cardiomyocytes in this xenogenic setting, an immunosuppressive regimen with cyclosporine and methylprednisolone was used in combination with an inhibitory antibody targeting CTLA4. Intramyocardial injection of one billion hESC cardiomyocytes resulted in substantial graft sizes of organized and cross-striated cardiomyocytes and functional coupling to the host myocardium via electrome-chanical junctions; however, functional recovery of the infarcted monkey heart could not be shown in this first study [206]. The crab-eating macaque (*Macaca fascicularis* or cynomolgus monkey) is also suitable as a preclinical animal model because of the anatomic and physiologic similarities to the human heart. The coronary system with the left anterior descending artery and left circumflex is similar, and vessel occlusion can be used for the induction of myocardial infarction. An adult cynomolgus monkey weighs 4–11 kg, but their heart-to-body weight ratio of ~0.5% is similar to that of humans [207]. Although their heart rate is higher (100–135 bpm) [208, 209], these primates are frequently used to evaluate drug-induced arrhythmias [210] and as a relevant preclinical animal model to investigate the consequences of and treatment options for myocardial infarction [211]. In contrast to the pig, well-established ESCs have been described for the cynomolgus monkey [212]. These cells are truly pluripotent and can differentiate into cells of all three germ layers [213, 214], including contracting cardiomyocytes, and are very similar to cynomolgus monkey iPSC stem cells [215]. The same holds true for iPSCs from rhesus macaques (*Macaca mulatta*), which can be differentiated into atrial-like and ventricular cardiomyocytes [216]. Therefore, nonhuman primate iPSCs offer a unique opportunity for use as an allogenic [217] or potentially even autologous transplantation model for evaluating the therapeutic potential of EHT.

3 Conclusions and Future Perspectives

In conclusion, a number of requirements must be fulfilled for the clinical application of in vitro engineered human myocardial tissue. We have discussed these requirements in detail and summarized them in Table 1. In addition, the functionality and therapeutic potential of engineered human heart tissue must be assessed in suitable animal models. To date, it is still unknown whether recovery of myocardial function can be better achieved with cell or tissue transplantation, or which of these therapeutic concepts can provide the optimal treatment strategy for a specific disease condition. Therefore, both concepts should be taken to (pre)clinical testing to find out how stem cell-derived myocardial cells can be used in clinical applications for the treatment of cardiovascular diseases.

Acknowledgments We thank Sylvia Merkert for help with TALEN-mediated targeted transgene integration into the AAVS1 locus and Anke Gawol for the generation of clonal iPSC lines.

References

1. WHO (2011) Cardiovascular diseases (CVDs). Fact Sheet No 317. World Health Organization, Geneva. Available at http://www.who.int/mediacentre/factsheets/fs317/en/index.html
2. Nadal-Ginard B (1978) Commitment, fusion and biochemical differentiation of a myogenic cell line in the absence of DNA synthesis. Cell 15(3):855–864
3. Soonpaa MH, Field LJ (1998) Survey of studies examining mammalian cardiomyocyte DNA synthesis. Circ Res 83(1):15–26
4. Bergmann O et al (2009) Evidence for cardiomyocyte renewal in humans. Science 324 (5923):98–102
5. Leor J et al (2000) Bioengineered cardiac grafts: a new approach to repair the infarcted myocardium? Circulation 102(19 Suppl 3):III56–III61
6. Zimmermann WH et al (2000) Three-dimensional engineered heart tissue from neonatal rat cardiac myocytes. Biotechnol Bioeng 68(1):106–114
7. Zimmermann WH, Tiburcy M, Eschenhagen T (2007) Cardiac tissue engineering: a clinical perspective. Futur Cardiol 3(4):435–445
8. Kreutziger KL, Murry CE (2011) Engineered human cardiac tissue. Pediatr Cardiol 32 (3):334–341
9. Nichols M et al. (2012) European cardiovascular disease statistics 2012. European Heart Network, Brussels, European Society of Cardiology, Sophia Antipolis
10. Boldt LH, Haverkamp W (2009) Arrhythmogenic right ventricular cardiomyopathy: diagnosis and risk stratification. Herz 34(4):290–297
11. Burnside N, MacGowan SW (2012) Malignant primary cardiac tumours. Interact Cardiovasc Thorac Surg 15(6):1004–1006
12. Leja MJ, Shah DJ, Reardon MJ (2011) Primary cardiac tumors. Tex Heart Inst J 38 (3):261–262
13. Stiver K et al (2015) Left atrial myxoma causing coronary steal: an atypical cause of angina. Tex Heart Inst J 42(3):270–272
14. Reynen K (1995) Cardiac myxomas. N Engl J Med 333(24):1610–1617
15. Ceresoli GL et al (1997) Primary cardiac lymphoma in immunocompetent patients: diagnostic and therapeutic management. Cancer 80(8):1497–1506

16. Penny DJ, Vick Iii GW (2011) Ventricular septal defect. Lancet 377(9771):1103–1112
17. Scully BB et al (2010) Current expectations for surgical repair of isolated ventricular septal defects. Ann Thorac Surg 89(2):544–551
18. Noonan JA, Nadas AS (1958) The hypoplastic left heart syndrome; an analysis of 101 cases. Pediatr Clin N Am 5(4):1029–1056
19. Mair R (2010) Aortenatresie, hypoplastisches Linksherzsyndrom und hypoplastischer Linksherzkomplex. Herzchirurgie. Springer, Berlin, pp 461–472
20. Tchervenkov CI et al (1998) Biventricular repair in neonates with hypoplastic left heart complex. Ann Thorac Surg 66(4):1350–1357
21. Eschenhagen T et al (1997) Three-dimensional reconstitution of embryonic cardiomyocytes in a collagen matrix: a new heart muscle model system. FASEB J 11(8):683–694
22. Fink C et al (2000) Chronic stretch of engineered heart tissue induces hypertrophy and functional improvement. FASEB J 14(5):669–679
23. Kofidis T et al (2002) In vitro engineering of heart muscle: artificial myocardial tissue. J Thorac Cardiovasc Surg 124(1):63–69
24. Zimmermann WH et al (2002) Tissue engineering of a differentiated cardiac muscle construct. Circ Res 90(2):223–230
25. Kensah G et al (2011) A novel miniaturized multimodal bioreactor for continuous in situ assessment of bioartificial cardiac tissue during stimulation and maturation. Tissue Eng Part C Methods 17(4):463–473
26. Hansen A et al (2010) Development of a drug screening platform based on engineered heart tissue. Circ Res 107(1):35–44
27. Naito H et al (2006) Optimizing engineered heart tissue for therapeutic applications as surrogate heart muscle. Circulation 114(1 Suppl):I72–I78
28. Schueller PO et al (2007) Intracoronary autologous bone marrow cell transplantation beneficially modulates heart rate variability. Int J Cardiol 119(3):398–399
29. Orlic D et al (2001) Bone marrow cells regenerate infarcted myocardium. Nature 410 (6829):701–705
30. Nygren JM et al (2004) Bone marrow-derived hematopoietic cells generate cardiomyocytes at a low frequency through cell fusion, but not transdifferentiation. Nat Med 10(5):494–501
31. Strauer BE et al (2002) Repair of infarcted myocardium by autologous intracoronary mononuclear bone marrow cell transplantation in humans. Circulation 106(15):1913–1918
32. Meyer GP et al (2009) Intracoronary bone marrow cell transfer after myocardial infarction: 5-year follow-up from the randomized-controlled BOOST trial. Eur Heart J 30 (24):2978–2984
33. Schaefer A et al (2010) Long-term effects of intracoronary bone marrow cell transfer on diastolic function in patients after acute myocardial infarction: 5-year results from the randomized-controlled BOOST trial--an echocardiographic study. Eur J Echocardiogr 11 (2):165–171
34. Gnecchi M et al (2008) Paracrine mechanisms in adult stem cell signaling and therapy. Circ Res 103(11):1204–1219
35. Korf-Klingebiel M et al (2015) Myeloid-derived growth factor (C19orf10) mediates cardiac repair following myocardial infarction. Nat Med 21(2):140–149
36. Pijnappels DA et al (2008) Forced alignment of mesenchymal stem cells undergoing cardiomyogenic differentiation affects functional integration with cardiomyocyte cultures. Circ Res 103(2):167–176
37. Rose RA et al (2008) Bone marrow-derived mesenchymal stromal cells express cardiac-specific markers, retain the stromal phenotype, and do not become functional cardiomyocytes in vitro. Stem Cells 26(11):2884–2892
38. Gruh I, Martin U (2009) Transdifferentiation of stem cells: a critical view. Adv Biochem Eng Biotechnol 114:73–106
39. Urbanek K et al (2003) Intense myocyte formation from cardiac stem cells in human cardiac hypertrophy. Proc Natl Acad Sci U S A 100(18):10440–10445

40. Beltrami AP et al (2003) Adult cardiac stem cells are multipotent and support myocardial regeneration. Cell 114(6):763–776
41. Bearzi C et al (2007) Human cardiac stem cells. Proc Natl Acad Sci U S A 104 (35):14068–14073
42. Mishra R et al (2011) Characterization and functionality of cardiac progenitor cells in congenital heart patients. Circulation 123(4):364–373
43. Sato H et al (2007) Detection of TUNEL-positive cardiomyocytes and c-kit-positive progenitor cells in children with congenital heart disease. J Mol Cell Cardiol 43(3):254–261
44. Emmert MY et al (2012) Higher frequencies of BCRP+ cardiac resident cells in ischaemic human myocardium. Eur Heart J 34(36):2830–2838
45. Oh H et al (2003) Cardiac progenitor cells from adult myocardium: homing, differentiation, and fusion after infarction. Proc Natl Acad Sci U S A 100(21):12313–12318
46. van Vliet P et al (2008) Progenitor cells isolated from the human heart: a potential cell source for regenerative therapy. Neth Heart J 16(5):163–169
47. Weissman IL, Anderson DJ, Gage F (2001) Stem and progenitor cells: origins, phenotypes, lineage commitments, and transdifferentiations. Annu Rev Cell Dev Biol 17:387–403
48. Bolli R et al (2011) Cardiac stem cells in patients with ischaemic cardiomyopathy (SCIPIO): initial results of a randomised phase 1 trial. Lancet 378(9806):1847–1857
49. van Berlo JH et al (2014) c-kit+ cells minimally contribute cardiomyocytes to the heart. Nature 509(7500):337–341
50. Keith MC, Bolli R (2015) "String theory" of c-kit(pos) cardiac cells: a new paradigm regarding the nature of these cells that may reconcile apparently discrepant results. Circ Res 116(7):1216–1230
51. Martin CM et al (2004) Persistent expression of the ATP-binding cassette transporter, Abcg2, identifies cardiac SP cells in the developing and adult heart. Dev Biol 265(1):262–275
52. Pfister O et al (2008) Role of the ATP-binding cassette transporter Abcg2 in the phenotype and function of cardiac side population cells. Circ Res 103(8):825–835
53. Scharenberg CW, Harkey MA, Torok-Storb B (2002) The ABCG2 transporter is an efficient Hoechst 33342 efflux pump and is preferentially expressed by immature human hematopoietic progenitors. Blood 99(2):507–512
54. Zhou S et al (2001) The ABC transporter Bcrp1/ABCG2 is expressed in a wide variety of stem cells and is a molecular determinant of the side-population phenotype. Nat Med 7 (9):1028–1034
55. Oyama T et al (2007) Cardiac side population cells have a potential to migrate and differentiate into cardiomyocytes in vitro and in vivo. J Cell Biol 176(3):329–341
56. Messina E et al (2004) Isolation and expansion of adult cardiac stem cells from human and murine heart. Circ Res 95(9):911–921
57. Smith RR et al (2007) Regenerative potential of cardiosphere-derived cells expanded from percutaneous endomyocardial biopsy specimens. Circulation 115(7):896–908
58. Lee ST et al (2011) Intramyocardial injection of autologous cardiospheres or cardiosphere-derived cells preserves function and minimizes adverse ventricular remodeling in pigs with heart failure post-myocardial infarction. J Am Coll Cardiol 57(4):455–465
59. Malliaras K et al (2014) Intracoronary cardiosphere-derived cells after myocardial infarction: evidence of therapeutic regeneration in the final 1-year results of the CADUCEUS trial (CArdiosphere-derived aUtologous stem CElls to reverse ventricUlar dySfunction). J Am Coll Cardiol 63(2):110–122
60. Chimenti I et al (2010) Relative roles of direct regeneration versus paracrine effects of human cardiosphere-derived cells transplanted into infarcted mice. Circ Res 106(5):971–980
61. Xu Y et al (2014) Cardiac differentiation of cardiosphere-derived cells in scaffolds mimicking morphology of the cardiac extracellular matrix. Acta Biomater 10(8):3449–3462
62. Li Z et al (2016) Thermosensitive and highly flexible hydrogels capable of stimulating cardiac differentiation of cardiosphere-derived cells under static and dynamic mechanical training conditions. ACS Appl Mater Interfaces 8(25):15948–15957

63. Kim DH et al (2012) Nanopatterned cardiac cell patches promote stem cell niche formation and myocardial regeneration. Integr Biol 4(9):1019–1033
64. Chimenti I et al (2011) Human cardiosphere-seeded gelatin and collagen scaffolds as cardiogenic engineered bioconstructs. Biomaterials 32(35):9271–9281
65. Thomson JA et al (1998) Embryonic stem cell lines derived from human blastocysts. Science 282(5391):1145–1147
66. Takahashi K et al (2007) Induction of pluripotent stem cells from adult human fibroblasts by defined factors. Cell 131(5):861–872
67. Yu J et al (2007) Induced pluripotent stem cell lines derived from human somatic cells. Science 318(5858):1917–1920
68. Haase A et al (2009) Generation of induced pluripotent stem cells from human cord blood. Cell Stem Cell 5(4):434–441
69. Streckfuss-Bömeke K et al (2012) Comparative study of human-induced pluripotent stem cells derived from bone marrow cells, hair keratinocytes, and skin fibroblasts. Eur Heart J 34 (33):2618–2629
70. Narazaki G et al (2008) Directed and systematic differentiation of cardiovascular cells from mouse induced pluripotent stem cells. Circulation 118(5):498–506
71. Mauritz C et al (2008) Generation of functional murine cardiac myocytes from induced pluripotent stem cells. Circulation 118(5):507–517
72. Schenke-Layland K et al (2008) Reprogrammed mouse fibroblasts differentiate into cells of the cardiovascular and hematopoietic lineages. Stem Cells 26(6):1537–1546
73. Kim JB et al (2008) Pluripotent stem cells induced from adult neural stem cells by reprogramming with two factors. Nature 454(7204):646
74. Di Stefano B, Prigione A, Broccoli V (2009) Efficient genetic reprogramming of unmodified somatic neural progenitors uncovers the essential requirement of Oct4 and Klf4. Stem Cells Dev 18(5):707–716
75. Kim JB et al (2009) Oct4-induced pluripotency in adult neural stem cells. Cell 136 (3):411–419
76. Huangfu D et al (2008) Induction of pluripotent stem cells by defined factors is greatly improved by small-molecule compounds. Nat Biotechnol 26(7):795–797
77. Shi Y et al (2008) Induction of pluripotent stem cells from mouse embryonic fibroblasts by Oct4 and Klf4 with small-molecule compounds. Cell Stem Cell 3(5):568–574
78. Wernig M et al (2008) A drug-inducible transgenic system for direct reprogramming of multiple somatic cell types. Nat Biotechnol 26(8):916–924
79. Maherali N et al (2008) A high-efficiency system for the generation and study of human induced pluripotent stem cells. Cell Stem Cell 3(3):340–345
80. Stadtfeld M et al (2008) Induced pluripotent stem cells generated without viral integration. Science 322(5903):945–949
81. Fusaki N et al (2009) Efficient induction of transgene-free human pluripotent stem cells using a vector based on Sendai virus, an RNA virus that does not integrate into the host genome. Proc Jpn Acad Ser B Phys Biol Sci 85(8):348–362
82. Macarthur CC et al (2012) Generation of human-induced pluripotent stem cells by a nonintegrating RNA Sendai virus vector in feeder-free or xeno-free conditions. Stem Cells Int 2012:564612
83. Okita K et al (2008) Generation of mouse induced pluripotent stem cells without viral vectors. Science 322(5903):949–953
84. Kaji K et al (2009) Virus-free induction of pluripotency and subsequent excision of reprogramming factors. Nature 458(7239):771
85. Ban H et al (2011) Efficient generation of transgene-free human induced pluripotent stem cells (iPSCs) by temperature-sensitive Sendai virus vectors. Proc Natl Acad Sci U S A 108 (34):14234–14239
86. Hockemeyer D et al (2009) Efficient targeting of expressed and silent genes in human ESCs and iPSCs using zinc-finger nucleases. Nat Biotechnol 27(9):851–857

87. Hockemeyer D et al (2011) Genetic engineering of human pluripotent cells using TALE nucleases. Nat Biotechnol 29(8):731–734

88. Maggio I et al (2014) Adenoviral vector delivery of RNA-guided CRISPR/Cas9 nuclease complexes induces targeted mutagenesis in a diverse array of human cells. Sci Rep 4:5105

89. Merkert S et al (2014) Efficient designer nuclease-based homologous recombination enables direct PCR screening for footprintless targeted human pluripotent stem cells. Stem Cell Rep 2 (1):107–118

90. Merkert S, Martin U (2017) Targeted gene editing in human pluripotent stem cells using site-specific nucleases. Adv Biochem Eng/Biotechnol. https://doi.org/10.1007/10_2017_XX

91. Mummery C (2011) Induced pluripotent stem cells--a cautionary note. N Engl J Med 364 (22):2160–2162

92. Ronen D, Benvenisty N (2012) Genomic stability in reprogramming. Curr Opin Genet Dev 22(5):444–449

93. Schwanke K et al (2006) Generation and characterization of functional cardiomyocytes from rhesus monkey embryonic stem cells. Stem Cells 24(6):1423–1432

94. Passier R et al (2005) Increased cardiomyocyte differentiation from human embryonic stem cells in serum-free cultures. Stem Cells 23(6):772–780

95. Dai W et al (2007) Survival and maturation of human embryonic stem cell-derived cardiomyocytes in rat hearts. J Mol Cell Cardiol 43(4):504–516

96. Graichen R et al (2008) Enhanced cardiomyogenesis of human embryonic stem cells by a small molecular inhibitor of p38 MAPK. Differentiation 76(4):357–370

97. Xu XQ et al (2008) Chemically defined medium supporting cardiomyocyte differentiation of human embryonic stem cells. Differentiation 76(9):958–970

98. Burridge PW et al (2012) Production of de novo cardiomyocytes: human pluripotent stem cell differentiation and direct reprogramming. Cell Stem Cell 10(1):16–28

99. Lian X et al (2012) Robust cardiomyocyte differentiation from human pluripotent stem cells via temporal modulation of canonical Wnt signaling. Proc Natl Acad Sci U S A 109(27): E1848–E1857

100. Lian X et al (2013) Directed cardiomyocyte differentiation from human pluripotent stem cells by modulating Wnt/beta-catenin signaling under fully defined conditions. Nat Protoc 8 (1):162–175

101. Zhu WZ et al (2010) Neuregulin/ErbB signaling regulates cardiac subtype specification in differentiating human embryonic stem cells. Circ Res 107(6):776–786

102. Ben-Ari M et al (2016) Developmental changes in electrophysiological characteristics of human-induced pluripotent stem cell-derived cardiomyocytes. Heart Rhythm 13 (12):2379–2387

103. Zhang J et al (2012) Extracellular matrix promotes highly efficient cardiac differentiation of human pluripotent stem cells: the matrix sandwich method. Circ Res 111(9):1125–1136

104. Burridge PW et al (2014) Chemically defined generation of human cardiomyocytes. Nat Methods 11(8):855–860

105. Kempf H et al (2014) Controlling expansion and cardiomyogenic differentiation of human pluripotent stem cells in scalable suspension culture. Stem Cell Rep 3(6):1132–1146

106. Kempf H, Zweigerdt R (2017) Adv Biochem Engin/Biotechnol. doi: https://doi.org/10.1007/ 10_2017_XX

107. Kleger A et al (2010) Modulation of calcium-activated potassium channels induces cardiogenesis of pluripotent stem cells and enrichment of pacemaker-like cells. Circulation 122(18):1823–1836

108. Jara-Avaca M et al (2017) EBIO does not induce cardiomyogenesis in human pluripotent stem cells but modulates cardiac subtype enrichment by lineage-selective survival. Stem Cell Rep 8(2):305–317

109. Zhang Q et al (2011) Direct differentiation of atrial and ventricular myocytes from human embryonic stem cells by alternating retinoid signals. Cell Res 21(4):579–587

110. Devalla HD et al (2015) Atrial-like cardiomyocytes from human pluripotent stem cells are a robust preclinical model for assessing atrial-selective pharmacology. EMBO Mol Med 7 (4):394–410
111. Protze SI et al (2017) Sinoatrial node cardiomyocytes derived from human pluripotent cells function as a biological pacemaker. Nat Biotechnol 35(1):56–68
112. Guo XM et al (2006) Creation of engineered cardiac tissue in vitro from mouse embryonic stem cells. Circulation 113(18):2229–2237
113. Wang X et al (2006) Scalable producing embryoid bodies by rotary cell culture system and constructing engineered cardiac tissue with ES-derived cardiomyocytes in vitro. Biotechnol Prog 22(3):811–818
114. Dubois NC et al (2011) SIRPA is a specific cell-surface marker for isolating cardiomyocytes derived from human pluripotent stem cells. Nat Biotechnol 29(11):1011–1018
115. Uosaki H et al (2011) Efficient and scalable purification of cardiomyocytes from human embryonic and induced pluripotent stem cells by VCAM1 surface expression. PLoS One 6 (8):e23657
116. Tohyama S et al (2013) Distinct metabolic flow enables large-scale purification of mouse and human pluripotent stem cell-derived cardiomyocytes. Cell Stem Cell 12(1):127–137
117. Hattori F, Fukuda K (2012) A method for purifying cardiomyocytes or programmed cardiomyocytes derived from stem cells. European Patent 1983042 B1
118. Hattori F, Fukuda K (2012) Method for inducing cell death in pluripotent stem cells and differentiated cells other than cardiac myocytes. European Patent 2415862 A1
119. Hattori F et al (2010) Nongenetic method for purifying stem cell-derived cardiomyocytes. Nat Methods 7(1):61–66
120. Klug MG et al (1996) Genetically selected cardiomyocytes from differentiating embronic stem cells form stable intracardiac grafts. J Clin Investig 98(1):216–224
121. Zandstra PW et al (2003) Scalable production of embryonic stem cell-derived cardiomyocytes. Tissue Eng 9(4):767–778
122. Xu XQ et al (2008) Highly enriched cardiomyocytes from human embryonic stem cells. Cytotherapy 10(4):376–389
123. Shimko VF, Claycomb WC (2008) Effect of mechanical loading on three-dimensional cultures of embryonic stem cell-derived cardiomyocytes. Tissue Eng Part A 14(1):49–58
124. Pfannkuche K et al (2010) Fibroblasts facilitate the engraftment of embryonic stem cell-derived cardiomyocytes on three-dimensional collagen matrices and aggregation in hanging drops. Stem Cells Dev 19(10):1589–1599
125. Liau B et al (2011) Pluripotent stem cell-derived cardiac tissue patch with advanced structure and function. Biomaterials 32(35):9180–9187
126. Kensah G et al (2013) Murine and human pluripotent stem cell-derived cardiac bodies form contractile myocardial tissue in vitro. Eur Heart J 34(15):1134–1146
127. Miki K et al (2015) Efficient detection and purification of cell populations using synthetic MicroRNA switches. Cell Stem Cell 16(6):699–711
128. Kuang Y et al (2017) Efficient, selective removal of human pluripotent stem cells via Ecto-alkaline phosphatase-mediated aggregation of synthetic peptides. Cell Chem Biol 24 (6):685–694 e4
129. Caspi O et al (2007) Tissue engineering of vascularized cardiac muscle from human embryonic stem cells. Circ Res 100(2):263–272
130. Lesman A et al (2010) Transplantation of a tissue-engineered human vascularized cardiac muscle. Tissue Eng Part A 16(1):115–125
131. Shapira-Schweitzer K et al (2009) A photopolymerizable hydrogel for 3-D culture of human embryonic stem cell-derived cardiomyocytes and rat neonatal cardiac cells. J Mol Cell Cardiol 46(2):213–224
132. Chen QZ et al (2010) An elastomeric patch derived from poly(glycerol sebacate) for delivery of embryonic stem cells to the heart. Biomaterials 31(14):3885–3893

133. Stevens KR et al (2009) Scaffold-free human cardiac tissue patch created from embryonic stem cells. Tissue Eng Part A 15(6):1211–1222
134. Stevens KR et al (2009) Physiological function and transplantation of scaffold-free and vascularized human cardiac muscle tissue. Proc Natl Acad Sci U S A 106(39):16568–16573
135. Matsuura K et al (2011) Creation of mouse embryonic stem cell-derived cardiac cell sheets. Biomaterials 32(30):7355–7362
136. Kwon OH et al (2000) Rapid cell sheet detachment from poly(N-isopropylacrylamide)-grafted porous cell culture membranes. J Biomed Mater Res 50(1):82–89
137. Schaaf S et al (2011) Human engineered heart tissue as a versatile tool in basic research and preclinical toxicology. PLoS One 6(10):e26397
138. Breckwoldt K et al (2017) Differentiation of cardiomyocytes and generation of human engineered heart tissue. Nat Protoc 12(6):1177–1197
139. Tiburcy M et al (2017) Defined engineered human myocardium with advanced maturation for applications in heart failure modeling and repair. Circulation 135(19):1832–1847
140. Larsen R (2009) Überwachung der Herz-Kreislauf-Funktion. Anästhesie und Intensivmedizin in Herz-, Thorax- und Gefäßchirurgie. Springer, Berlin, pp 51–69
141. Hoppe UC, Erdmann E (2011) Chronische Herzinsuffizienz. In: Erdmann E (ed) Klinische Kardiologie. Springer, Berlin, pp 123–179
142. Willruth AM et al (2011) Comparison of global and regional right and left ventricular longitudinal peak systolic strain, strain rate and velocity in healthy fetuses using a novel feature tracking technique. J Perinat Med 39(5):549–556
143. Perles Z et al (2007) Assessment of fetal myocardial performance using myocardial deformation analysis. Am J Cardiol 99(7):993–996
144. Wiegerinck RF et al (2009) Force frequency relationship of the human ventricle increases during early postnatal development. Pediatr Res 65(4):414–419
145. Holubarsch C et al (1998) Shortening versus isometric contractions in isolated human failing and non-failing left ventricular myocardium: dependency of external work and force on muscle length, heart rate and inotropic stimulation. Cardiovasc Res 37(1):46–57
146. Mulieri LA et al (1992) Altered myocardial force-frequency relation in human heart failure. Circulation 85(5):1743–1750
147. Sarsero D et al (2003) (−)-CGP 12177 increases contractile force and hastens relaxation of human myocardial preparations through a propranolol-resistant state of the beta 1-adrenoceptor. Naunyn Schmiedeberg's Arch Pharmacol 367(1):10–21
148. Holubarsch C et al (1996) Existence of the Frank-Starling mechanism in the failing human heart. Investigations on the organ, tissue, and sarcomere levels. Circulation 94(4):683–689
149. Siedner S et al (2003) Developmental changes in contractility and sarcomeric proteins from the early embryonic to the adult stage in the mouse heart. J Physiol 548(2):493–505
150. Black 3rd LD et al (2009) Cell-induced alignment augments twitch force in fibrin gel-based engineered myocardium via gap junction modification. Tissue Eng Part A 15(10):3099–3108
151. Engelmayr Jr GC et al (2008) Accordion-like honeycombs for tissue engineering of cardiac anisotropy. Nat Mater 7(12):1003–1010
152. Tulloch NL et al (2011) Growth of engineered human myocardium with mechanical loading and vascular coculture. Circ Res 109(1):47–59
153. Zhang D et al (2013) Tissue-engineered cardiac patch for advanced functional maturation of human ESC-derived cardiomyocytes. Biomaterials 34(23):5813–5820
154. Jackman CP, Carlson AL, Bursac N (2016) Dynamic culture yields engineered myocardium with near-adult functional output. Biomaterials 111:66–79
155. Mannhardt I et al (2016) Human engineered heart tissue: analysis of contractile force. Stem Cell Rep 7(1):29–42
156. Seta H et al (2017) Tubular cardiac tissues derived from human induced pluripotent stem cells generate pulse pressure in vivo. Sci Rep 7:45499
157. Nunes SS et al (2013) Biowire: a platform for maturation of human pluripotent stem cell-derived cardiomyocytes. Nat Methods 10(8):781–787

158. Murry CE, Reinecke H, Pabon LM (2006) Regeneration gaps: observations on stem cells and cardiac repair. J Am Coll Cardiol 47(9):1777–1785
159. Kim JS et al (2008) Correlation of serial cardiac magnetic resonance imaging parameters with early resolution of ST-segment elevation after primary percutaneous coronary intervention. Circ J 72(10):1621–1626
160. Masugata H et al (1999) Relationship between myocardial tissue density measured by microgravimetry and sound speed measured by acoustic microscopy. Ultrasound Med Biol 25(9):1459–1463
161. Kawel N et al (2012) Normal left ventricular myocardial thickness for middle-aged and older subjects with steady-state free precession cardiac magnetic resonance: the multi-ethnic study of atherosclerosis. Circ Cardiovasc Imaging 5(4):500–508
162. Dahlmann J et al (2013) The use of agarose microwells for scalable embryoid body formation and cardiac differentiation of human and murine pluripotent stem cells. Biomaterials 34 (10):2463–2471
163. Zweigerdt R et al (2011) Scalable expansion of human pluripotent stem cells in suspension culture. Nat Protoc 6(5):689–700
164. Olmer R et al (2012) Suspension culture of human pluripotent stem cells in controlled, stirred bioreactors. Tissue Eng Part C Methods 18(10):772–784
165. Radisic M et al (2006) Oxygen gradients correlate with cell density and cell viability in engineered cardiac tissue. Biotechnol Bioeng 93(2):332–343
166. Shimizu T et al (2006) Polysurgery of cell sheet grafts overcomes diffusion limits to produce thick, vascularized myocardial tissues. FASEB J 20(6):708–710
167. Masumoto H et al (2012) Pluripotent stem cell-engineered cell sheets reassembled with defined cardiovascular populations ameliorate reduction in infarct heart function through cardiomyocyte-mediated neovascularization. Stem Cells 30(6):1196–1205
168. Gao L et al (2017) Myocardial tissue engineering with cells derived from human-induced pluripotent stem cells and a native-like, high-resolution, 3-dimensionally printed scaffold. Circ Res 120(8):1318–1325
169. Nakane T et al (2017) Impact of cell composition and geometry on human induced pluripotent stem cells-derived engineered cardiac tissue. Sci Rep 7:45641
170. Zhang B et al (2016) Biodegradable scaffold with built-in vasculature for organ-on-a-chip engineering and direct surgical anastomosis. Nat Mater 15(6):669–678
171. Komae H et al (2017) Three-dimensional functional human myocardial tissues fabricated from induced pluripotent stem cells. J Tissue Eng Regen Med 11(3):926–935
172. Togo S et al (2011) Differentiation of embryonic stem cells into fibroblast-like cells in three-dimensional type I collagen gel cultures. In Vitro Cell Dev Biol Anim 47(2):114–124
173. Hewitt KJ et al (2011) Epigenetic and phenotypic profile of fibroblasts derived from induced pluripotent stem cells. PLoS One 6(2):e17128
174. Lin B et al (2012) High-purity enrichment of functional cardiovascular cells from human iPS cells. Cardiovasc Res 95(3):327–335
175. Dar A et al (2012) Multipotent vasculogenic pericytes from human pluripotent stem cells promote recovery of murine ischemic limb. Circulation 125(1):87–99
176. Zimmermann WH et al (2006) Engineered heart tissue grafts improve systolic and diastolic function in infarcted rat hearts. Nat Med 12(4):452–458
177. Singla DK et al (2011) Induced pluripotent stem (iPS) cells repair and regenerate infarcted myocardium. Mol Pharm 8(5):1573–1581
178. Rojas SV et al (2017) Transplantation of purified iPSC-derived cardiomyocytes in myocardial infarction. PLoS One 12(5):e0173222
179. van Laake LW et al (2008) Human embryonic stem cell-derived cardiomyocytes and cardiac repair in rodents. Circ Res 102(9):1008–1010
180. Mummery CL, Davis RP, Krieger JE (2010) Challenges in using stem cells for cardiac repair. Sci Transl Med 2(27):27ps17

181. Garbern JC, Mummery CL, Lee RT (2013) Model systems for cardiovascular regenerative biology. Cold Spring Harb Perspect Med 3(4):a014019
182. Riegler J et al (2015) Human engineered heart muscles engraft and survive long term in a rodent myocardial infarction model. Circ Res 117(8):720–730
183. Shiba Y et al (2012) Human ES-cell-derived cardiomyocytes electrically couple and suppress arrhythmias in injured hearts. Nature 489(7415):322–325
184. Shiba Y et al (2014) Electrical integration of human embryonic stem cell-derived cardiomyocytes in a Guinea pig chronic infarct model. J Cardiovasc Pharmacol Ther 19 (4):368–381
185. Weinberger F et al (2016) Cardiac repair in guinea pigs with human engineered heart tissue from induced pluripotent stem cells. Sci Transl Med 8(363):363ra148
186. Williams AR et al (2013) Enhanced effect of combining human cardiac stem cells and bone marrow mesenchymal stem cells to reduce infarct size and to restore cardiac function after myocardial infarction. Circulation 127(2):213–223
187. Okura H et al (2012) Intracoronary artery transplantation of cardiomyoblast-like cells from human adipose tissue-derived multi-lineage progenitor cells improve left ventricular dysfunction and survival in a swine model of chronic myocardial infarction. Biochem Biophys Res Commun 425(4):859–865
188. Emmert MY et al (2013) Transcatheter based electromechanical mapping guided intramyocardial transplantation and in vivo tracking of human stem cell based three dimensional microtissues in the porcine heart. Biomaterials 34(10):2428–2441
189. Kim BO et al (2005) Cell transplantation improves ventricular function after a myocardial infarction: a preclinical study of human unrestricted somatic stem cells in a porcine model. Circulation 112(9 Suppl):I96–104
190. Moelker AD et al (2007) Intracoronary delivery of umbilical cord blood derived unrestricted somatic stem cells is not suitable to improve LV function after myocardial infarction in swine. J Mol Cell Cardiol 42(4):735–745
191. Ghodsizad A et al (2009) Transplanted human cord blood-derived unrestricted somatic stem cells improve left-ventricular function and prevent left-ventricular dilation and scar formation after acute myocardial infarction. Heart 95(1):27–35
192. Gahremanpour A et al (2013) Xenotransplantation of human unrestricted somatic stem cells in a pig model of acute myocardial infarction. Xenotransplantation 20(2):110–122
193. Templin C et al (2012) Transplantation and tracking of human-induced pluripotent stem cells in a pig model of myocardial infarction: assessment of cell survival, engraftment, and distribution by hybrid single photon emission computed tomography/computed tomography of sodium iodide symporter transgene expression. Circulation 126(4):430–439
194. Xiong Q et al (2011) A fibrin patch-based enhanced delivery of human embryonic stem cell-derived vascular cell transplantation in a porcine model of postinfarction left ventricular remodeling. Stem Cells 29(2):367–375
195. Xiong Q et al (2013) Functional consequences of human induced pluripotent stem cell therapy: myocardial ATP turnover rate in the in vivo swine heart with postinfarction remodeling. Circulation 127(9):997–1008
196. Kawamura M et al (2012) Feasibility, safety, and therapeutic efficacy of human induced pluripotent stem cell-derived cardiomyocyte sheets in a porcine ischemic cardiomyopathy model. Circulation 126(11 Suppl 1):S29–S37
197. Roberts RM et al (2015) Livestock models for exploiting the promise of pluripotent stem cells. ILAR J 56(1):74–82
198. Wu Z et al (2009) Generation of pig induced pluripotent stem cells with a drug-inducible system. J Mol Cell Biol 1(1):46–54
199. Ezashi T et al (2009) Derivation of induced pluripotent stem cells from pig somatic cells. Proc Natl Acad Sci U S A 106(27):10993–10998
200. Esteban MA et al (2009) Generation of induced pluripotent stem cell lines from Tibetan miniature pig. J Biol Chem 284(26):17634–17640

201. Kues WA et al (2013) Derivation and characterization of sleeping beauty transposon-mediated porcine induced pluripotent stem cells. Stem Cells Dev 22(1):124–135
202. Gandolfi F et al (2012) Why is it so difficult to derive pluripotent stem cells in domestic ungulates? Reprod Domest Anim 47(Suppl 5):11–17
203. Linke A et al (2005) Stem cells in the dog heart are self-renewing, clonogenic, and multipotent and regenerate infarcted myocardium, improving cardiac function. Proc Natl Acad Sci U S A 102(25):8966–8971
204. Yarbrough WM, Spinale FG (2003) Large animal models of congestive heart failure: a critical step in translating basic observations into clinical applications. J Nucl Cardiol 10 (1):77–86
205. Shimada H et al (2010) Generation of canine induced pluripotent stem cells by retroviral transduction and chemical inhibitors. Mol Reprod Dev 77(1):2
206. Chong JJ et al (2014) Human embryonic-stem-cell-derived cardiomyocytes regenerate non-human primate hearts. Nature 510(7504):273–277
207. Keenan CM, Vidal JD (2006) Standard morphologic evaluation of the heart in the laboratory dog and monkey. Toxicol Pathol 34(1):67–74
208. Malinow MR, Hill JD, Ochsner 3rd AJ (1977) Heart rate in caged Macaca Fascicularis. Effects of short-term physical exercise. J Med Primatol 6(2):69–75
209. Kaplan JR, Manuck SB, Gatsonis C (1990) Heart rate and social status among male cynomolgus monkeys (Macaca Fascicularis) housed in disrupted social groupings. Am J Primatol 21(3):175–187
210. Misner DL et al (2012) Investigation of mechanism of drug-induced cardiac injury and torsades de pointes in cynomolgus monkeys. Br J Pharmacol 165(8):2771–2786
211. Yang XM et al (2010) Attenuation of infarction in cynomolgus monkeys: preconditioning and postconditioning. Basic Res Cardiol 105(1):119–128
212. Suemori H et al (2001) Establishment of embryonic stem cell lines from cynomolgus monkey blastocysts produced by IVF or ICSI. Dev Dyn 222(2):273–279
213. Akama K et al (2011) Proteomic identification of differentially expressed genes in neural stem cells and neurons differentiated from embryonic stem cells of cynomolgus monkey (Macaca Fascicularis) in vitro. Biochim Biophys Acta (BBA) Proteins Proteomics 1814 (2):265–276
214. Kobayashi M et al (2008) BMP4 induces primitive endoderm but not trophectoderm in monkey embryonic stem cells. Cloning Stem Cells 10(4):495–502
215. Wunderlich S et al (2012) Induction of pluripotent stem cells from a cynomolgus monkey using a polycistronic simian immunodeficiency virus-based vector, differentiation toward functional cardiomyocytes, and generation of stably expressing reporter lines. Cell Reprogram 14(6):471–484
216. Zhang X et al (2017) Differentiation and characterization of rhesus monkey atrial and ventricular cardiomyocytes from induced pluripotent stem cells. Stem Cell Res 20:21–29
217. Shiba Y et al (2016) Allogeneic transplantation of iPS cell-derived cardiomyocytes regenerates primate hearts. Nature 538(7625):388–391

Adv Biochem Eng Biotechnol (2018) 163: 147–168
DOI: 10.1007/10_2017_28
© Springer International Publishing AG 2017
Published online: 1 November 2017

Human Pluripotent Stem Cells to Engineer Blood Vessels

Xin Yi Chan, Morgan B. Elliott, Bria Macklin, and Sharon Gerecht

Abstract Development of pluripotent stem cells (PSCs) is a remarkable scientific advancement that allows scientists to harness the power of regenerative medicine for potential treatment of disease using unaffected cells. PSCs provide a unique opportunity to study and combat cardiovascular diseases, which continue to claim the lives of thousands each day. Here, we discuss the differentiation of PSCs into vascular cells, investigation of the functional capabilities of the derived cells, and their utilization to engineer microvascular beds or vascular grafts for clinical application.

X.Y. Chan, M.B. Elliott, B. Macklin, and S. Gerecht (✉)
Department of Chemical and Biomolecular Engineering, Institute for NanoBioTechnology,
Johns Hopkins University, Baltimore, MD 21218, USA
e-mail: gerecht@jhu.edu

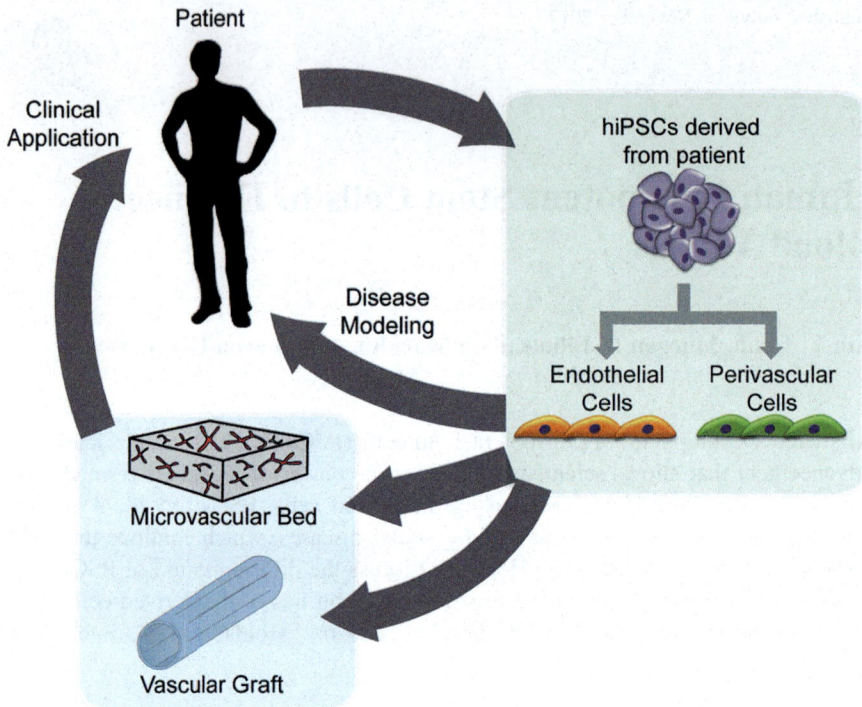

Graphical Abstract Human iPSCs generated from patients are differentiated toward ECs and perivascular cells for use in disease modeling, microvascular bed development, or vascular graft fabrication

Keywords Human pluripotent stem cells, Small-diameter tissue engineered vascular grafts, Vascular differentiation, Vascular disease modeling, Vascular networks

Contents

Abbreviations

2D	Two-dimensional
3D	Three-dimensional
bFGF	Basic fibroblast growth factor
BMPR2	Bone morphogenetic protein receptor type II
BP	Burst pressure
CAD	Coronary artery disease
CCD	Chronic cardiovascular defects
DO	Dissolved oxygen
DPI	Diphenyleneiodonium
EB	Embryoid body
EC	Endothelial cell
ECM	Extracellular matrix
EVC	Early vascular cell
FBN1	Fibrillin1
FPAH	Family members of pulmonary arterial hypertension
HA	Hyaluronic acid
(h)ESC	(Human) embryonic stem cell
HIF	Hypoxia-inducible factors
(h)[i]PSC	(Human) [induced] pluripotent stem cell
HUVECs	Human umbilical vein endothelial cells
ITA	Internal thoracic artery
MFS	Marfan syndrome
MMP	Matrix metalloproteinase
PDGF-BB	Platelet-derived growth factor-BB
PEG	Poly(ethylene glycol)
PEGDA	PEG-diacrylate
PGA	Polyglycolic acid
ROS	Reactive oxygen species
SMA	Smooth muscle actin
SMMHC	Smooth muscle myosin heavy chain
SRS	Suture retention strength
(s)TEVG	(Small-diameter) tissue engineered vascular graft
SV	Saphenous vein
TESA	Tissue engineering by self-assembly
TGFβ	Transforming growth factor β
UMC	Unaffected mutation carrier
VEGF(R)	Vascular endothelial growth factor (receptor)
vSMC	Vascular smooth muscle cell

1   Introduction

Functional blood vessels are essential for delivering oxygen and nutrients.[1] These vessels are specialized and can be categorized into several classes, including arteries, arterioles, capillaries, venules, and veins. A blood vessel consists of a tube lined with endothelial cells (ECs) in the inner wall and surrounded by support cells such as pericytes and vascular smooth muscle cells (vSMCs), depending on the vascular size (Fig. 1). ECs play an important role as a barrier to pathogens and in many physiological processes such as wound healing, the immuno/inflammatory response, and coagulation [2].

Typically, vascular disease occurs when the cellular makeup of the patient's vasculature changes. Causes of vascular diseases are often linked to genetic disorders such as peripheral arterial hypertension. Historically, transgenic animal models of mice and zebrafish have played an important role in modeling cardiovascular diseases, characterizing the pathology and physiology of the disease, identifying downstream targets, and evaluating therapeutic drugs and treatments. However, following an increased number of promising drug treatment failures in clinical trials, the use of animal models in testing new therapeutic drugs has been criticized for its ineffectiveness. In recent decades, with the development of human induced pluripotent stem cells (hiPSCs) as a source of patient-specific regenerative therapies, hiPSCs have become an ideal alternative to animal models or patient tissue samples as a platform for modeling vascular diseases, because they carry the same genetic abnormalities as the patients from whom the cells were derived. These in vitro hiPSC-generated vascular disease models could possibly advance medical treatment by providing mechanistic insights into vascular diseases and discovering

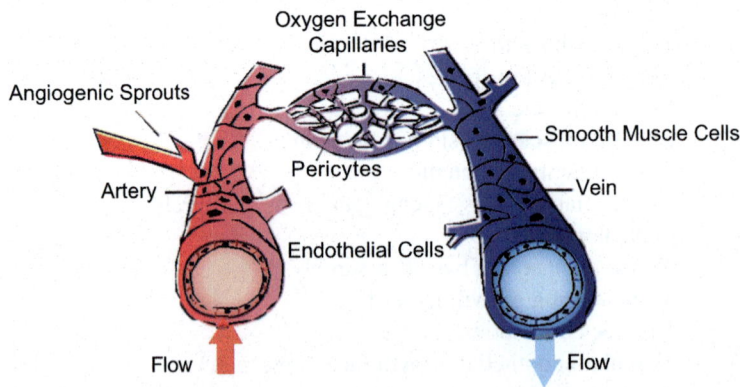

Fig. 1 Cellular makeup of blood vessels. Taken from [1]

---

[1]Note: Alterations to the root abbreviation are indicated in parentheses or brackets.

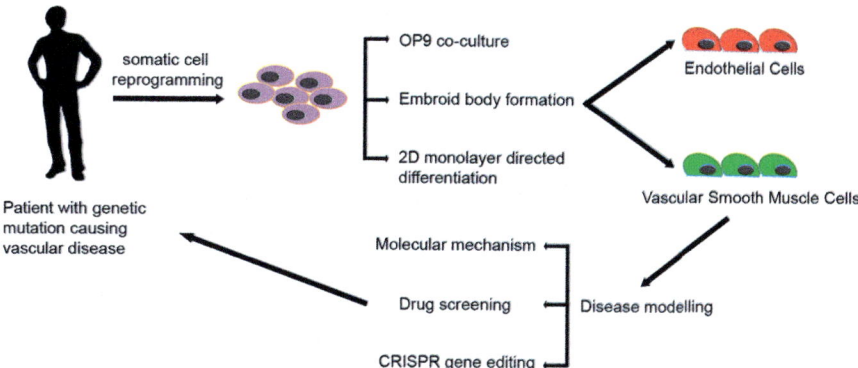

Fig. 2 Modeling genetic mutations in vascular diseases using hiPSC-derived ECs and vSMCs

new drugs via large-scale drug screens. Here, we discuss the different approaches used to derive vascular cells from healthy hiPSCs and some examples of vascular diseases modeled using hiPSC-derived vascular cells (Fig. 2).

2 Derivation of ECs and Perivascular Cells from Healthy and Diseased hiPSCs

2.1 EC and vSMC Differentiation from hPSCs

The first instance of reprogramming human somatic cells into hiPSCs was reported by two independent research groups, those of Yamanaka and Thompson. Yamanaka's group showed that using retroviruses to transfect four factors (OCT4, KLF4, SOX2, and C-MYC) into human fibroblasts is sufficient to reprogram those cells into hiPSCs [3]. On the other hand, Thomson's group demonstrated that reprogramming hiPSCs can be achieved using lentiviral vectors to transfect a different set of factors, including OCT4, NANOG, LIN28, and SOX2 [4]. Generation of hiPSCs from different cell sources without integration of reprogramming factors into the genome can improve the quality of these cells without posing potential risks of a genome-integrating virus vector backbone [5, 6].

Based on published work in vertebrates such as rodents and zebrafish, developmental factors and chemical molecules were utilized to guide hiPSCs to differentiate and mature into functional vascular derivatives. Over the last few decades, multiple protocols have been established to derive vascular cells from hPSCs, including both ECs and vSMCs. In general, the approaches for vascular differentiation described here were first demonstrated with human embryonic stem cells (hESCs) and then similar methods and their improvements were developed for hiPSCs.

2.1.1 OP9 Co-culture

The OP9 cell line is a stromal cell line derived from the skullcap of mice with an osteopetrotic mutation in the gene encoding macrophage colony-stimulating factor, a factor that has inhibitory effects on hematopoietic differentiation. Vodyanik et al. demonstrated that hPSCs cultured on an OP9 feeder layer can be directed to differentiate into a subset of CD34$^+$ hematopoietic progenitors and then matured into CD31$^+$CD34$^+$CD43$^-$ ECs [7]. They discovered that differentiation of hPSCs on top of the OP9 stromal cells is sufficient to generate a large number or CD34$^+$ cells without adding cytokines. This result indicates the importance of paracrine signaling and cytokines secreted by OP9 cells to direct the differentiation of hematopoietic and endothelial lineages [8]. After improving OP9 co-culture differentiation, another group of researchers found that CD31$^+$CD34$^+$ vascular progenitor cells differentiated on OP9 feeder cells could be further differentiated and matured into functional ECs and vSMCs separately when cultured in specific media supporting their specific lineage differentiation [9]. The EC derivatives express CD31, CD144, VEGRF2, and CD105, whereas the vSMC derivatives express desmin, α-smooth muscle actin (αSMA), calponin, and SM22α.

2.1.2 Three-Dimensional Differentiation Via Embryoid Body Formation

Another technique utilized to differentiate ECs and vSMCs is embryoid body (EB) formation. EB aggregates mimic primitive streak formation and induction of all three germ layers during embryonic development by responding to similar cues [10]. Also, because of their nonadherent nature, EB aggregates can be cultured in suspension, which is scalable and thus enables large production of differentiated ECs [11].

EC differentiation: Many endothelial differentiation methods typically grow hPSCs into EBs and culture them in suspension in differentiation media for a few days. Differentiation steps typically involve cell sorting (either magnetic or fluorescence-activated cell sorting; FACS) to isolate vascular progenitor cells using markers such as CD34, CD31, and CD144, followed by their culture with specific small molecules and growth factors to promote differentiation and maturation to hematopoietic and endothelial lineages [12, 13]. Different combinations of growth factors have been used in different methods. For example, James et al. added BMP4, activin A, and basic fibroblast growth factor (bFGF) to hPSC-derived EB aggregates to initiate differentiation. Thereafter, they transferred the EBs onto Matrigel and added vascular endothelial growth factor (VEGF)-A and a small molecule inhibitor of the transforming growth factor β (TGFβ) signaling pathway [14], thereby increasing the yield of CD31$^+$ ECs tenfold.

vSMC differentiation: Similar to endothelial differentiation, a variety of methods with different combinations of growth factors and extracellular membrane proteins have been utilized to guide differentiation into vSMCs. The main difference is in the cellular and molecular markers used to validate and assess the successful derivation of vSMCs [15, 16]. In a method developed by Lin et al. [86], hiPSC EBs were treated with growth factors such as VEGF-A and bFGF during early differentiation to direct the differentiation of multipotent cardiovascular progenitor cells. These KDR^{low}c-kit$^-$ progenitor cells were sorted using FACS and subsequently cultured as a monolayer with VEGF and bFGF added to the medium. A second sort was performed to isolate cells that were $CD31^-CD166^-$ to further direct differentiation into functional vSMCs utilizing a specific smooth muscle growth medium.

2.1.3   2D Monolayer Differentiation

Directed differentiation as a two-dimensional (2D) monolayer of hPSCs is another method for guiding vascular differentiation, aiming to overcome limitations such as the relative heterogeneous differentiation of EBs, which could be a result of limited diffusion of chemical cues to the interior of the EBs. In addition, monolayer differentiation methods guide lineage commitment and can increase cell yield and viability after sorting.

EC differentiation: Using a 2D culture, scientists can fine tune the chemical cues necessary to induce EC fate directly by adding growth factors and small molecules. Using this approach, differentiation from hPSCs was optimized to a content of 50–70% ECs prior to sorting [17–20]. In our recently published protocols, Kusuma et al. [19] and Chan et al. [20] demonstrated the hiPSC-based derivation of early vascular cells (EVCs), which are characterized by the expression of vascular endothelial cadherin and platelet-derived growth factor receptor β. EVCs can be matured into ECs or pericytes and, when encapsulated in a synthetic hydrogel, can interact with each other, undergo morphogenesis, and self-organize into 3D vascular networks.

vSMC differentiation: There are several differentiation protocols demonstrating successful derivation of vSMCs from hiPSCs using growth factors and extracellular matrix (ECM) proteins to guide the differentiation. However, to date, only two protocols reliably differentiate hiPSCs into vSMCs with either synthetic or contractile lineage specification [21, 22]. The synthetic phenotype is characterized by high proliferation, migration, and ECM protein production. The contractile phenotype is characterized by low proliferation, low synthetic activity, and expression of contractile proteins, namely, smooth muscle myosin heavy chain (SMMHC) and elastin. Based on our protocol described by Wanjare et al. [21], hiPSCs were seeded on collagen IV in the first stage of differentiation to derive mesodermal cells and then, following addition of platelet derived growth factor-BB (PDGF-BB) and TGFβ with 10% serum, were derived into synthetic vSMCs. These vSMC derivatives express αSMA, calponin, and SM22a; about 50% also express SMMHC. Continuous culture of these cells in TGFβ and low serum medium can further mature the cells into contractile vSMCs, which express SMMHC and elastin.

2.2 Human iPSCs as a Tool to Model Vascular Diseases

The development of iPSC technology has opened up avenues for study of vascular disease by overcoming the challenges of species-specific limitations resulting from animal models. In addition, the difficulty of harvesting sufficient patient vascular tissue samples can be overcome by deriving these tissues from hiPSCs generated from the patient. Both ECs and vSMCs play a crucial role in maintaining vascular function. Genetic mutations affecting development of the vasculature can result in dysfunctional vasculature, leading to vascular diseases such as pulmonary hypertension, Marfan syndrome (MFS), or others (outlined in Sects. 2.2.1 and 2.2.2). The ability to differentiate vascular cell types from hiPSCs enables researchers to study the molecular and pathophysiology aspects of these diseases. In addition, the stages of differentiation of these vascular cells from hiPSCs closely mimic their developmental stages in vivo, presenting a unique opportunity to model and study disease progression in vitro.

2.2.1 Human iPSC-EC Disease Modeling

Recently, patient-specific iPSC-ECs have been employed to study pathways involved in pulmonary arterial hypertension (PAH). In PAH, dysfunctional ECs of the pulmonary arteries are the key factor in the initiation and progression of the disease. These dysfunctional ECs in PAH display phenotypes showing features such as decreased cell survival upon injury, impaired adhesion and migration, and disordered angiogenesis. Gu, Shao and colleagues generated patient-specific iPSC-ECs from family members of pulmonary arterial hypertension (FPAH) patients and unaffected mutation carriers (UMC) of bone morphogenetic protein receptor type II (BMPR2) mutation, and compared them with gender-matched controls to investigate the protective modifiers of the BMPR2 mutation [23]. The group demonstrated that EC morphology and BMPR2 expression are similar in FPAH and UMC iPSC-ECs. However, FPAH iPSC-ECs had impaired cell adhesion on multiple ECM substrates and reduced cell survival after serum withdrawal. Elevated BMPR2 activators and reduced BMPR2 inhibitors in UMC iPSC-ECs are responsible for the BMPR2-mediated activation of p-P38 signaling and increased β1-integrin, which improve cell adhesion. Independent of the BMPR2 pathway, the authors also discovered that increased levels of baculoviral IAP repeat-containing 3 (BIRC3) in the UMC iPSC-ECs improved cell survival. Furthermore, correction of the BMPR2 mutation using CRISPR restored the functions of rescued FPAH-ECs to those of control iPSC-ECs. These findings shed light on the importance of protective modifiers for FPAH, which could help in developing potential treatments for FPAH.

2.2.2 Human iPSC-vSMC Disease Modeling

Marfan syndrome (MFS) is a heritable genetic disorder caused by mutations in fibrillin1 (FBN1) that affect the connective tissue of patients due to dysfunctional vSMCs. Patients with this disease often have vSMC defects that affect FBN1 accumulation, ECM degradation, TGFβ signaling, and contraction and apoptosis of the vSMCs. The Sinha group successfully generated MFS-vSMCs from patient-specific MFS-iPSCs [24]. According to the authors, MFS-vSMCs exhibited the same symptoms as in the aortas of Marfan patients. These cells have reduced levels of FBN1 deposition and increased levels of TGFβ and matrix metalloproteinases (MMPs) in the ECM. In addition, MFS-vSMCs showed functional abnormalities, including higher incidence of cell death associated with increased vSMC loss in MFS aortic dilatation, and reduced contractility similar to that observed in MFS aortas. The abnormalities in MFS-vSMCs can be rescued by correction of the FBN1 mutation. Inhibition of TGFβ in MFS-vSMCs was sufficient to rescue the phenotypes of FBN1 reduction and MMP increase, but not to alleviate the high incidence of cell apoptosis, which is regulated by the non-canonical p38 pathway. This particular vSMC disease model provides a platform to study the molecular mechanisms affecting MFS and help develop future therapeutic approaches.

3 Harnessing the Extracellular Cues to Engineer Vascular Networks from hiPSCs

3.1 Angiogenesis and Vasculogenesis

Vasculogenesis and angiogenesis are the primary processes that regulate blood vessel formation in all blooded species. Vasculogenesis is the de novo formation of vascular structures, whereas angiogenesis is the formation of vessels from preexisting vasculature. Vasculogenesis occurs in three developmental steps, beginning with cells of the mesoderm. These early mesodermal cells first differentiate into blood islands, which are bicellular aggregates comprising angioblasts on the outer layer and hemopoietic cells internally. Next, in response to an increase in growth factor binding to VEGFR2, VEGFR1, and tie-1, the angioblasts differentiate into ECs. Newly created ECs form the primary vascular plexus, an embryonic structure from which all subsequent vessels form via angiogenesis [25]. Angiogenesis begins with a specific EC, referred to as the "tip cell," which is activated by cues in the embryonic environment and then leads the sprouting process. Stalk cells, which are in direct contact with the tip cell, begin to proliferate and form laminated structures. Mural cells, including pericytes and vSMCs, are recruited to stabilize the newly formed vasculature by EC-derived ligands (i.e., heparin-binding epidermal growth factor and PDGF-BB) [26].

These complex processes can be mimicked using 3D vascular models in vitro. These models include a natural or synthetic biomaterial-based scaffold to serve as the ECM for the networks, several growth factors that can be added to induce the process, and other physical cues such as oxygen and matrix stiffeners to simulate the surroundings during vasculogenesis and angiogenesis.

3.2 Biomaterials

To engineer a viable 3D model, the biomaterial selected must be optimal for the intended purpose. As new vasculature is created, ECs must constantly remodel their ECM via traction forces, proteolytic activity, and cell–matrix adhesion to allow sprouting and lumen formation [27]. Biomaterials for use in vascular network models must allow this remodeling, in addition to being biocompatible and possessing optimal stiffness, structure, and permeability. Hydrogels are materials that have a high water content, yet do not dissolve in water. Hydrogels simulate natural tissues in that they can retain structural integrity in highly aqueous environments and allow easy diffusion of small molecules. Natural, synthetic, and semisynthetic hydrogels are widely used to recapitulate vasculogenesis and angiogenesis.

Commonly used natural hydrogels include collagen, fibrin, and gelatin. These proteins are produced naturally within the body and, thus, the hydrogel derivatives are characteristically biocompatible and biodegradable. Type I collagen gel, for example, allows EC network formation through activation of tubulogenesis pathways (Fig. 3). EC sprouting and migration occur through the creation of "vascular guidance tunnels" via α2β1 integrin binding and MT1-MMP network degradation. When supporting mural cells are introduced, matrix remodeling and ECM

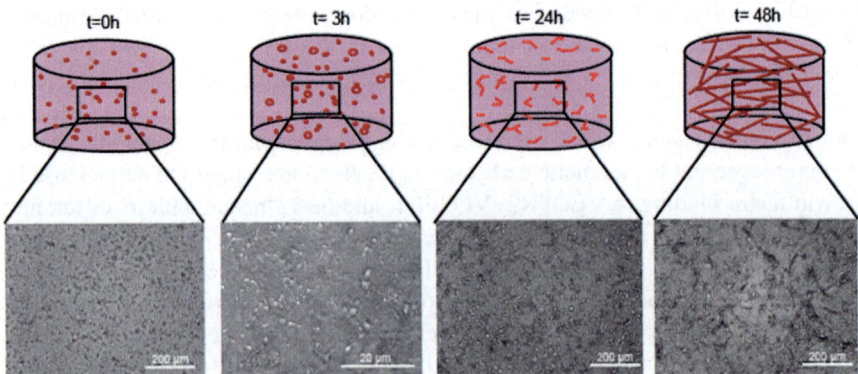

Fig. 3 Vascular assembly kinetics. Schematic (*upper panel*) and corresponding light microscopy images (*lower panel*) showing the progression of vascular assembly of iPSC-derived ECs encapsulated in collagen gels ($t = 0$ h), including vacuole formation ($t = 3$ h), sprouting events ($t = 24$ h), and network growth ($t = 48$ h). Graphics not drawn to scale. Taken from [28]

production increases, resulting in α5β1, α3β1, α6β1, and α1β1 integrin binding [29]. Collagen gel is also an optimal biomaterial for vascular modeling because of the high concentration of collagen in the body and its natural load-bearing capabilities [30]. Synthetic hydrogels are highly amendable and can be custom-made to the desired structure, stiffness, and degradability. This allows the creation of a much more defined and tunable system. Although natural materials can vary in uniformity from batch to batch, synthetic hydrogels maintain consistency between batches. A commonly used material for synthetic hydrogels is poly(ethylene glycol) (PEG). One major drawback of using synthetic hydrogels such as PEG is that they have little or no cell adhesion or degradation sites. This shortcoming can be resolved by adding functional sites to the polymer, including adhesive sites such as arginine-glycine-aspartic acid (RGD) sequences or MMP-degradable sites, which are routinely utilized to improve cell–material interaction. Semisynthetic hydrogels are a new class of biomaterials that incorporate the advantages of natural materials with the customizability of synthetic polymers. Examples include acrylated hyaluronic acid (HA) and dextran hydrogels [31].

3.2.1 Matrix Properties

Stiffness: The stiffness of a biomaterial is typically determined by the elastic modulus. The stiffness of natural materials such as collagen and fibrin hydrogels can be modified by increasing the density, which has been shown to affect neovessel growth and sprouting [32, 33]. More specifically, the modulus of hydrogel materials can be altered by increasing the polymer concentration or changing the crosslinking density of the material. We have previously demonstrated the effects of stiffness on endothelial progenitor cells using a semisynthetic HA–gelatin hydrogel with PEG-diacrylate (PEGDA) as crosslinker. By modifying the concentration of PEGDA to 1, 0.4, and 0.1%, three significantly different Young's moduli were generated, creating rigid, firm, and yielding hydrogels, respectively. Physical and biological analyses of the networks affirmed the crucial role of matrix stiffness. Cell cultures in the yielding hydrogel possessed a significantly higher mean tube length, tube area, and tube thickness than cultures in the rigid and stiff hydrogels. Both firm and yielding substrates allowed formation of luminal structures, whereas the stiff hydrogel did not. In response to high concentrations of VEGF, yielding hydrogels showed a decreased expression of MT1-MMP, MMP-1, and MMP-2 [34].

Degradation: As mentioned above, the ability of ECs to degrade their ECM is paramount to both angiogenesis and vasculogenesis. Sokic and Papavasiliou utilized a PEGDA-based hydrogel to demonstrate this [35]. Hydrogels were made using MMP-sensitive peptides with either one or three proteolytic cleavage sites and functionalized to PEGDA macromeres with one or multiple MMP-sensitive peptide domains between each crosslink. Hydrogels with only one MMP cleavage site took up to 96 h to degrade completely, depending upon the weight percentage used, whereas hydrogels with three cleavage sites degraded in as little as 1 h. The authors used human umbilical vein endothelial cells (HUVECs) in an invasion assay to show that hydrogels with more cleavage sites had a greater invasion area and depth of invasion [35]. The ability to control degradation of the hydrogel has also allowed creation of gels that permit controlled release of various growth factors. This is an efficient and directed approach to deliver growth factors. Heprasil™, a hybrid mesh of poly(ε--caprolactone)-collagen blend and HA hydrogel, was dual-loaded with VEGF and PDGF-BB, which were released over 21 days. During the 21 days, HUVECs continued to grow in response to the growth factors [36].

Through optimization of both adhesion and degradation sites, as well as stiffness of acrylated HA hydrogels, we have shown the activation of vasculogenesis pathways of endothelial progenitors [37]. More recently, we have shown that hiPSC-derived EVCs undergo tubulogenesis in these HA hydrogels, resulting in multicellular, functional vascular networks [19].

3.3 Oxygen and Hypoxia

Oxygen tension plays a key role in the regulation of angiogenesis and vasculogenesis, affecting cell viability, differentiation, migration, and ECM remodeling. Hypoxia-inducible factors (HIF) and reactive oxygen species (ROS) govern EC response and adaptation to changes in oxygen levels, allowing increases in crucial growth factors such as VEGF and bFGF [38, 39]. HIF1α controls angiogenesis in hypoxic oxygen levels of less than 1%. We have previously demonstrated the effects of hypoxia on HUVECs in a 3D collagen matrix. Hydrogels encapsulated with HUVECs were allowed to incubate for 48 h while oxygen partial pressure was measured. After 24 h, oxygen within the gel had decreased from ~12% to <5%, whereas gels supplemented with diphenyleneiodonium (DPI), a ROS inhibitor, had oxygen levels that remained at ~20%. Cell viability was also affected after 24 h, with a larger percentage of HUVECs dying in gels not supplemented with DPI. Gel thickness was also shown to regulate oxygen availability at the bottom of the gel. Although inhibition of ROS allowed greater cell viability, the vascular network characteristics (mean tube length, tube thickness, and tube area coverage) significantly decreased compared with untreated gels, providing overwhelming evidence for the importance of hypoxia in angiogenesis [40].

Although it has been shown that dissolved oxygen (DO) levels can also be regulated by adjusting the height of the hydrogel, this is highly uncontrolled and can vary when using natural materials. Our group was the first to synthesize a hydrogel that can regulate DO levels and gradients within its own 3D environment. We were able to show that DO levels can be precisely controlled by modifying reaction kinetics and hydrogel composition. An increase in vascular network characteristics was shown using these hypoxia-inducible hydrogels [41]. More recently, we have shown that EVCs derived from hiPSCs of both healthy and diabetic donors respond to the hypoxic environment, generating extensive vascular networks within the hypoxic hydrogels [20].

4 Vascular Graft Fabrication Using hiPSC Derivatives

4.1 Clinical Need

Tissue engineered vascular grafts (TEVGs) are in high demand for replacing harvested autologous vessels used as bypass, endovascular, and interposition grafts [42, 43]. Over 0.5 million patients with coronary artery disease (CAD) undergo coronary artery bypass procedures each year [43–46]. Meanwhile, 1% of children are born with chronic cardiovascular defects (CCD) and require repeated cardiac surgery to reconstruct vascular conduits [47–49]. For single ventricle cardiac anomalies, the most severe CCD, synthetic vascular grafts are the leading cause of complications resulting from their lack of growth during child development [49–53]. For both CAD and pediatric CCD cases, the standard treatment is to replace these small-diameter arteries with autologous tissue grafts [43, 49, 54–56], which have numerous disadvantages. Harvesting the tissue is inconvenient and there may be insufficient tissue available, limiting reconstruction [49, 54, 55]. Repeated surgery, multiple operation sites, limited availability, and sacrificed arteriovenous function to obtain a graft underscore the clinical need for a TEVG with the patency and low thrombogenicity that is characteristic of native vessel grafts.

Synthetic TEVGs are commonly used for procedures that require a graft larger than 6 mm in diameter [16]. However, for small-diameter TEVGs of under 6 mm in diameter (sTEVGs), synthetic materials have not shown clinical effectiveness and are inferior to autologous tissue grafts [49, 54, 55, 57]. Currently available artificial grafts have low durability because of atherosclerosis and stenosis and may catastrophically fail after 8–12 years [45]. Although several efforts using mature or progenitor cell lines have been successful in developing sTEVGs [58–62], a functional graft has remained elusive because of post-implantation challenges, including thrombogenicity, decreased elasticity, decreased compliance, aneurysmal failure, and intimal hyperplasia [58, 62, 63]. Significant improvement is required in order to provide CAD and pediatric CCD patients with an ideal sTEVG to replace the autologous graft gold standard.

4.1.1 The Ideal sTEVG

Patient-derived vessels should be matched to the size specifications of the patient and be able to grow with pediatric patients. The ideal engineered structure is nonimmunogenic, capable of scale-up with a clinically relevant shelf-life, has low thrombogenicity, and exhibits long-term patency [43, 60, 63]. Additionally, the sTEVG should have mechanical properties similar to those of native vessels such as the internal thoracic artery (ITA) and saphenous vein (SV), commonly used as autologous grafts [64]. The mechanical properties used to compare grafts with native tissue are burst pressure (BP), suture retention strength (SRS), and compliance. SV parameters have values of 2,134 mmHg, 1.92 N, and 25.6%/100 mmHg, respectively, and those for ITA are 3,073 mmHg, 1.72 N, and 11.5%/100 mmHg [64]. A fully biomimetic graft that recapitulates arterial properties is ideal and difficult to achieve [64].

The development of functional cellularized sTEVGs requires complex interactions and specific organization among ECs, vSMCs, and several ECM proteins. Crucially for sTEVG applications, ECs reduce platelet activation and adhesion, have antimicrobial properties, aid in fibrinolysis, and prevent intimal hyperplasia and leukocyte adhesion [65–67]. Meanwhile, vSMCs provide mechanical strength, vasoreactivity, and improved stability of TEVGs [68–70]. Although each cell type plays a unique role in vascular function, cellular crosstalk affects vessel function and further complicates graft fabrication [69, 71, 72]. Debate on the ideal cell source for vascular engineering is ongoing, but hiPSCs may be the answer for clinically relevant, patient-specific grafts [73–76]. The ability to derive ECs that can generate patient-specific blood vessels from type I diabetic patient-derived hiPSCs [20], a population with associated vascular diseases for which autologous vascular grafts may be difficult to obtain [73], shows the clinical relevance of the cell source and potential for relevant in vitro disease models. Design of a robust sTEVG seeded with hPSCs would provide a substantial benefit for patients.

4.2   Current Efforts to Develop sTEVGs

Within the field of vascular tissue engineering, there are three classes of techniques for developing sTEVGs: scaffold-based methods, tissue engineering by self-assembly (TESA), and decellularized matrices (Fig. 4) [64, 76]. Scaffold-based

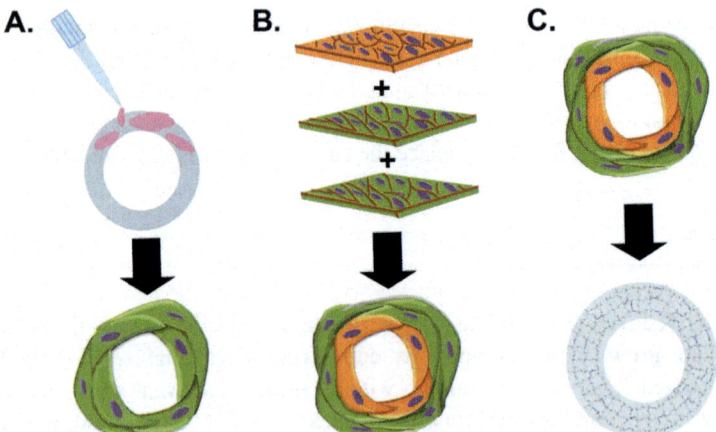

Fig. 4 Three techniques for fabricating sTEVGs. (**a**) A commonly used scaffold-based method is drip seeding a cell solution onto a graft-like structure that is typically made of synthetic material and vSMCs (*green*). (**b**) An efficient TESA method is culturing several types of cell sheets and concentrically rolling them to attain a multilayered graft structure. The cell sheets may be ECs (*orange*), vSMCs (*green*), or fibroblasts (not shown). (**c**) Native vessels or hPSC-derived engineered vessels can be decellularized to make an acellular natural matrix that can be used as an off-the-shelf vascular graft

methods focus on using a natural or synthetic matrix as a base, onto which cells are seeded [64]. TESA techniques do not utilize a scaffold or matrix [64]. The decellularization methods remove cells from vessels to fabricate an off-the-shelf, nonimmunogenic graft [64]. Using PSCs with these methods only started to gain momentum in 2012, but noteworthy progress has been made.

4.2.1 Scaffold-Based Methods

The Niklason laboratory pioneered scaffold-based methods for fabricating sTEVGs, originally beginning with mature vascular cells and recently adding hPSC-based approaches. The group's first attempt used hESC-derived mesenchymal stem cells that could be further differentiated using TGFβ1 into contractile calponin-positive vSMCs seeded onto a 1-mm diameter polyglycolic acid (PGA) scaffold [44]. After 8 weeks and applying pulsatile flow during graft culture, a collagen-rich cell wall that positively stained for αSMA was achieved [44]. The lack of graft calponin staining was unsurprising given the use of 20% serum for maintaining the synthetic vSMC phenotype and encouraging cell proliferation [44]. Subsequent reduction in serum concentration could lead to a contractile vSMC phenotype and increase calponin expression on the graft. The importance of the growth factor cocktail was highlighted by the combination of TGFβ1 for differentiation and bFGF for enhanced proliferation, which together with nutrient availability may have contributed to the expression of osteo- and chondrogenic markers near the lumen [44]. This hESC-derived attempt raises serious concerns over the vascular cell fate stability and plasticity on sTEVGs.

Niklson's group next focused on iPSCs drip-seeded on PGA scaffolds, splitting cultures into distinct proliferation (20% serum, PDGF-BB) and differentiation (10% serum, no growth factors) culture stages, each lasting 4 weeks [77]. The final stage incorporated mechanical stimulation to enhance differentiation [77]. Eliminating TGFβ1 and using only PDGF-BB reduced unwanted differentiation into osteo- and chondrogenic lineages [77]. This resulted in vSMCs positive for αSMA, SM22α, and calponin, but not the mature, contractile SMMHC marker [77]. It was again found that cells closer to the lumen were less differentiated, indicating that diffusion of nutrients or propagation of mechanical stimuli through the 250-μm thick wall, similar to the SV, may have had an effect [77]. The structure was highly collagenous, containing glycosaminoglycans and fibronectin, but no elastin [77], which is crucial for a biomimetic, mechanically responsive sTEVG. However, a BP of 700 mmHg and SRS of 30 g were measured for one graft [77]. Interestingly, karyotypically abnormal iPSCs led to high calcification and a senescent phenotype not seen with karyotypically normal cells [77]. Later, Gui et al. reverted to using both TGFβ1 and PDGF-BB with αSMA- and calponin-positive hiPSC-derived vSMCs to create a highly collagenous, SMMHC-positive sTEVG after being cultured for 9 weeks in vitro without mechanical stimulation [78]. Mature elastic fibers were still absent, despite a BP of 500 mmHg and SRS of 70 g in one graft [78]. Furthermore, the graft was implanted for 2 weeks as an abdominal aorta interposition graft in rats [78]. The graft did not rupture, remained patent, recruited host cells, and did not result in teratomas [78]. A longer in vivo study is necessary because maximum thrombus formation occurs over the first 4 weeks, teratomas may take 4–6 weeks to form, and the slight dilatation that occurred could increase over time [67, 78, 79]. These sTEVGs are some of the most mechanically robust, but also take the longest to develop, a significant barrier to clinical relevance.

Mechanical stimulation shortens the required culture period, as shown for hiPSC-derived ECs seeded in a bioreactor and stimulated with a shear stress of 5–10 dyn/cm^2, resulting in arterial-like mature cells within 24 h [80]. The latter scaffold-based case can be classified as a "pre-sTEVG" because the cylindrical poly (L-lactic acid) construct measuring 5 mm in diameter and seeded with iPSC-derived synthetic vSMCs would be a perfusable sTEVG if the center was punctured out to create tissue rings [76]. Using 5% FBS, Wang et al. were able to induce the contractile SMC phenotype [76]. A nonfunctional, 2-week in vivo study was performed 24 h after cell seeding, which resulted in collagenous matrix formation and maintenance of the SMC phenotype [76]. Although the mechanical properties of the pre-sTEVGs are not known, these two cases indicate that a more clinically relevant timeline could be attained for achieving graft functionality, especially with the combination of mechanical stimulation and decreased serum concentration.

4.2.2 Tissue Engineering by Self-Assembly

The first PSC self-assembly method used Matrigel encapsulation of mouse ESCs to form a cell layer on a four-well Labtek Chamber-Slide culture system and yielded varied displacement of the gel with shear stress [42]. Abilez et al. suggested that differentiation into ECs, SMCs, and fibroblasts could be used to create an autologous TEVG [42]. However, the group did not demonstrate the ability to differentiate ESCs and assemble a vessel from individual cell layers. The next attempt used iPSC-derived vascular cell sheets cultured on a temperature-responsive surface that were subsequently wrapped around a 0.8-mm diameter PGA–L-lactide and poly(L-lactide-co-ε-caprolactone) scaffold [73]. This seeding method increased cellularization efficiency by 80% relative to the drip seeding method used in scaffold-based methods [73]. Upon implantation in an inferior vena cava interposition model, the graft showed no thrombus or aneurysm formation, graft rupture, or calcification [73]. An abdominal aorta interposition model would be more rigorous because of increased pressure and shear stress; however, host cells had replaced the implanted cells by 10 weeks [73]. Alarmingly, iPSC-derived cells did not colocalize with vWF or αSMA, suggesting de-differentiation. Furthermore, 25% of mice had teratomas [73]. Improved cell lineage commitment, complete differentiation, and cellular purification are needed before implanting PSC-seeded sTEVGs.

In a unique self-assembly method, a ring shaped agarose well containing culture medium including 20% FBS, PDGF-BB, and TGFβ1 was used to form highly cellularized, uniformly thick vascular conduits from hiPSC-derived vSMCs [81]. After 14 days of culture, tissue rings of 2 mm inner diameter and robust, contractile, and highly collagenous walls of 0.84 mm thickness were attained

[81]. Although αSMA, SM22α, calponin, SMMHC, and elastin markers were present after differentiation, the presence of elastin and mature elastic fibers within the rings was not examined [81]. In combination with a 21-day differentiation protocol [81], this was one of the most clinically relevant timelines for facile production of robust sTEVGs. The group also modeled supravalvular stenosis syndrome by producing rings with decreased contractility, decreased SMMHC expression, and increased proliferation [81]. This study yielded both healthy and diseased physiologically relevant sTEVGs.

4.2.3 Decellularized Matrices

Combining this sTEVG fabrication method with PSCs has been minimally investigated, but any of the mentioned efforts could be included by adding a decellularization step. Carefully karyotyped and characterized iPSC-derived cells could yield sTEVGs with uniform biological and mechanical properties after decellularization [77]. Although the risks of immunogenicity and teratomas could be eliminated using acellular grafts of decellularized matrix, the exclusion of iPSC-derived vSMCs would decrease the mechanical properties of the graft, already below those of native vessels. Future efforts should investigate decellularized matrices from hPSC and hPSC derivatives as implantable, off-the-shelf sTEVGs.

4.3 Remaining Challenges

Opportunities for improving hiPSC-derived, patient-specific sTEVGs remain. Most sTEVGs would benefit from increased elastin content, better mechanical properties, lower cell lineage variability, reduced tumorigenesis, and clinically applicable production timelines. Culture of mature vSMCs and fibroblasts under pulsatile perfusion can yield mature elastic fibers within 30 days [82]. Both increased culture time and mechanical stimulation can increase elastin content of engineered vessels [82]. Similarly, increased ECM production can resolve the limited mechanical properties of PSC-derived sTEVGs because vessel mechanical properties are mostly provided by the ECM [82]. Co-culture of vSMCs with fibroblasts, which produce ECM significantly faster than vSMCs, might yield sTEVGs suitable for implantation [82]. Improved mechanical properties would help prevent dehiscence, rupture along the surgical anastomosis site, and compliance mismatch [76, 82].

The formation of teratomas shows the pressing need for the selectivity and maintenance of cell lineage commitment and fastidious purification of hPSC derivatives [44, 73]. However, hiPSCs from both healthy and type I diabetic patients can be reliably differentiated into vascular cells [83, 84]. A pure hPSC-derived population must be balanced with production of sTEVGs on a clinically applicable timeline, from cell isolation to graft cellularization. Samuel et al. found that 2D differentiation is more efficient than 3D differentiation and can be accomplished within 2 weeks [83]. Emergency situations could require the use of banked, patient-matched iPSC-derived sTEVGs or decellularized matrices [77], whereas more extended fabrication of patient-specific grafts would be possible for chronic cases. Producing xenogeneic-free sTEVGs using human serum may also be beneficial for clinical trials, because culture with 20% FBS can result in up to 30 mg of bovine serum proteins in cells, which could cause an immune reaction or zoonosis [85].

Most groups have only used PSC-derived vSMCs to fabricate sTEVGs, showing the infancy of the application of hPSCs to sTEVG construction. Co-culture of vSMCs with fibroblasts or ECs has been shown to be synergistic by increasing collagen production or reducing thrombus formation, respectively [76, 82]. Co-culture could yield a fully biomimetic sTEVG that can be used to treat CAD and pediatric CCD patients and provide opportunities to model human vascular diseases for investigation and drug testing, which may help prevent the failure of expensive clinical trials [81]. Ultimately, hPSC-derived vascular cells could provide the means to achieve the ideal, patient-specific sTEVG with low thrombogenicity, long-term patency, and mechanical properties similar to those of native vessels.

References

1. Smith Q, Gerecht S (2014) Going with the flow: microfluidic platforms in vascular tissue engineering. Curr Opin Chem Eng 3:42–50. https://doi.org/10.1016/j.coche.2013.11.001
2. Pate M, Damarla V., Chi, DS., Negi S, Krishnaswamy G. (2010) Adv Clin Chem 52:109–130
3. Takahashi K et al. (2007) Induction of pluripotent stem cells from adult human fibroblasts by defined factors. Cell 131:861–872. https://doi.org/10.1016/j.cell.2007.11.019
4. Yu J et al. (2007) Induced pluripotent stem cell lines derived from human somatic cells. Science 318:1917–1920. https://doi.org/10.1126/science.1151526
5. Li R et al. (2014) Shear stress–activated Wnt-Angiopoietin-2 signaling recapitulates vascular repair in zebrafish embryos. Arterioscler Thromb Vasc Biol 34:2268–2275
6. Deng XY et al. (2015) Non-viral methods for generating integration-free, induced pluripotent stem cells. Curr Stem Cell Res Ther 10:153–158
7. Choi KD et al. (2009) Hematopoietic and endothelial differentiation of human induced pluripotent stem cells. Stem Cells 27:559–567. https://doi.org/10.1634/stemcells.2008-0922
8. Vodyanik MA, Bork JA, Thomson JA, Slukvin II (2005) Human embryonic stem cell-derived CD34+ cells: efficient production in the coculture with OP9 stromal cells and analysis of lymphohematopoietic potential. Blood 105:617–626. https://doi.org/10.1182/blood-2004-04-1649
9. Marchand M et al. (2014) Concurrent generation of functional smooth muscle and endothelial cells via a vascular progenitor. Stem Cells Transl Med 3:91–97. https://doi.org/10.5966/sctm.2013-0124
10. Murry CE, Keller G (2008) Differentiation of embryonic stem cells to clinically relevant populations: lessons from embryonic development. Cell 132:661–680. https://doi.org/10.1016/j.cell.2008.02.008
11. Dang SM, Gerecht-Nir S, Chen J, Itskovitz-Eldor J, Zandstra PW (2004) Controlled, scalable embryonic stem cell differentiation culture. Stem Cells 22:275–282. https://doi.org/10.1634/stemcells.22-3-275

12. Rufaihah AJ et al. (2011) Endothelial cells derived from human iPSCS increase capillary density and improve perfusion in a mouse model of peripheral arterial disease. Arterioscler Thromb Vasc Biol 31:e72–e79. https://doi.org/10.1161/ATVBAHA.111.230938
13. Adams WJ et al. (2013) Functional vascular endothelium derived from human induced pluripotent stem cells. Stem Cell Rep 1:105–113. https://doi.org/10.1016/j.stemcr.2013.06.007
14. James D et al. (2010) Expansion and maintenance of human embryonic stem cell-derived endothelial cells by TGF beta inhibition is Id1 dependent. Nat Biotechnol 28:161–U115. https://doi.org/10.1038/nbt1605
15. Xie CQ et al. (2007) A highly efficient method to differentiate smooth muscle cells from human embryonic stem cells. Arterioscler Thromb Vasc Biol 27:e311–e312. https://doi.org/10.1161/ATVBAHA.107.154260
16. Lee TH et al. (2010) Functional recapitulation of smooth muscle cells via induced pluripotent stem cells from human aortic smooth muscle cells. Circ Res 106:120–128. https://doi.org/10.1161/CIRCRESAHA.109.207902
17. Patsch C et al. (2015) Generation of vascular endothelial and smooth muscle cells from human pluripotent stem cells. Nat Cell Biol 17:994–1003. https://doi.org/10.1038/ncb3205. http://www.nature.com/ncb/journal/v17/n8/abs/ncb3205.html - supplementary-information
18. Lian X et al. (2014) Efficient differentiation of human pluripotent stem cells to endothelial progenitors via small-molecule activation of WNT signaling. Stem Cell Rep 3:804–816. https://doi.org/10.1016/j.stemcr.2014.09.005
19. Kusuma S et al. (2013) Self-organized vascular networks from human pluripotent stem cells in a synthetic matrix. Proc Natl Acad Sci 110:12601–12606
20. Chan XY et al. (2015) Three-dimensional vascular network assembly from diabetic patient-derived induced pluripotent stem cells. Arterioscler Thromb Vasc Biol 35:2677–2685. https://doi.org/10.1161/ATVBAHA.115.306362
21. Wanjare M, Kuo F, Gerecht S (2013) Derivation and maturation of synthetic and contractile vascular smooth muscle cells from human pluripotent stem cells. Cardiovasc Res 97:321–330
22. Yang LB et al. (2016) Differentiation of human induced-pluripotent stem cells into smooth-muscle cells: two novel protocols. PLoS One 11(1):e0147155.ARTN e0147155. https://doi.org/10.1371/journal.pone.0147155
23. Gu M et al. (2016) Patient-specific iPSC-derived endothelial cells uncover pathways that protect against pulmonary hypertension in BMPR2 mutation carriers. Cell Stem Cell 20(4):490–504.e5. https://doi.org/10.1016/j.stem.2016.08.019
24. Granata A et al. (2017) An iPSC-derived vascular model of Marfan syndrome identifies key mediators of smooth muscle cell death. Nat Genet 49:97–109. https://doi.org/10.1038/ng.3723
25. Risau W, Flamme I (1995) Vasculogenesis. Annu Rev Cell Dev Biol 11:73–91. https://doi.org/10.1146/annurev.cb.11.110195.000445
26. Risau W (1997) Mechanisms of angiogenesis. Nature 386:671–674. https://doi.org/10.1038/386671a0
27. Edgar LT et al. (2014) Mechanical interaction of angiogenic microvessels with the extracellular matrix. J Biomech Eng 136:021001. https://doi.org/10.1115/1.4026471
28. Macklin BL, Gerecht S (2017) Bridging the gap: induced pluripotent stem cell derived endothelial cells for 3D vascular assembly. Curr Opin Chem Eng 15:102–109
29. Sacharidou A, Stratman AN, Davis GE (2012) Molecular mechanisms controlling vascular lumen formation in three-dimensional extracellular matrices. Cells Tissues Organs 195:122–143. https://doi.org/10.1159/000331410
30. Silver FH, Horvath I, Foran DJ (2001) Viscoelasticity of the vessel wall: the role of collagen and elastic fibers. Crit Rev Biomed Eng 29:279–301
31. Bersini S et al. (2016) Cell-microenvironment interactions and achritectures in microvascular systems. Biotechno Adv 34:1113–1130
32. Edgar LT, Underwood CJ, Guilkey JE, Hoying JB, Weiss JA (2014) Extracellular matrix density regulates the rate of neovessel growth and branching in sprouting angiogenesis. PLoS One 9:e85178. https://doi.org/10.1371/journal.pone.0085178

33. Park YK et al. (2014) In vitro microvessel growth and remodeling within a three-dimensional microfluidic environment. Cell Mol Bioeng 7:15–25. https://doi.org/10.1007/s12195-013-0315-6

34. Hanjaya-Putra D et al. (2010) Vascular endothelial growth factor and substrate mechanics regulate in vitro tubulogenesis of endothelial progenitor cells. J Cell Mol Med 14:2436–2447. https://doi.org/10.1111/j.1582-4934.2009.00981.x

35. Sokic S, Papavasiliou G (2012) Controlled proteolytic cleavage site presentation in biomimetic PEGDA hydrogels enhances neovascularization in vitro. Tissue Eng Part A 18:2477–2486. https://doi.org/10.1089/ten.TEA.2012.0173

36. Ekaputra AK, Prestwich GD, Cool SM, Hutmacher DW (2011) The three-dimensional vascularization of growth factor-releasing hybrid scaffold of poly (epsilon-caprolactone)/collagen fibers and hyaluronic acid hydrogel. Biomaterials 32:8108–8117. https://doi.org/10.1016/j.biomaterials.2011.07.022

37. Hanjaya-Putra D et al. (2011) Controlled activation of morphogenesis to generate a functional human microvasculature in a synthetic matrix. Blood 118:804–815. https://doi.org/10.1182/blood-2010-12-327338

38. Manalo DJ et al. (2005) Transcriptional regulation of vascular endothelial cell responses to hypoxia by HIF-1. Blood 105:659–669. https://doi.org/10.1182/blood-2004-07-2958

39. Calvani M, Rapisarda A, Uranchimeg B, Shoemaker RH, Melillo G (2006) Hypoxic induction of an HIF-1alpha-dependent bFGF autocrine loop drives angiogenesis in human endothelial cells. Blood 107:2705–2712. https://doi.org/10.1182/blood-2005-09-3541

40. Abaci HE, Truitt R, Tan S, Gerecht S (2011) Unforeseen decreases in dissolved oxygen levels affect tube formation kinetics in collagen gels. Am J Physiol Cell Physiol 301:C431–C440. https://doi.org/10.1152/ajpcell.00074.2011

41. Park KM, Gerecht S (2014) Hypoxia-inducible hydrogels. Nat Commun 5:4075. https://doi.org/10.1038/ncomms5075

42. Abilez O et al. (2006) A novel culture system shows that stem cells can be grown in 3D and under physiologic pulsatile conditions for tissue engineering of vascular grafts. J Surg Res 132:170–178

43. Gui L, Muto A, Chan SA, Breuer CK, Niklason LE (2009) Development of decellularized human umbilical arteries as small-diameter vascular grafts. Tissue Eng A 15:2665–2676

44. Sundaram S, Echter A, Sivarapatna A, Qiu C, Niklason L (2014) Small-diameter vascular graft engineered using human embryonic stem cell-derived mesenchymal cells. Tissue Eng A 20:740–750

45. Thompson CA et al. (2002) A novel pulsatile, laminar flow bioreactor for the development of tissue-engineered vascular structures. Tissue Eng 8:1083–1088

46. Mozaffarian D, Benjamin E, Go AS, Arnett DK, Blaha MJ, Cushman M, de Ferranti S, Després J-P, Fullerton HJ, Howard VJ, Huffman MD, Judd SE, Kissela BM, Lackland DT, Lichtman JH, Lisabeth LD, Liu S, Mackey RH, Matchar DB, DK MG, Mohler 3rd ER, Moy CS, Muntner P, Mussolino ME, Nasir K, Neumar RW, Nichol G, Palaniappan L, Pandey DK, Reeves MJ, Rodriguez CJ, Sorlie PD, Stein J, Towfighi A, Turan TN, Virani SS, Willey JZ, Woo D, Yeh RW, MBB T, on behalf of the American Heart Association Statistics Committee and Stroke Statistics Subcommittee (2015) Heart disease and stroke statistics—2015 update: a report from the American Heart Association. Circulation 131:e29–322. https://doi.org/10.1161/CIR.0000000000000152

47. Jia H, Caputo M, Ghorbel MT (2013) Stem cells in vascular graft tissue engineering for congenital heart surgery. Interv Cardiol 5:647–662

48. Children's Heart Foundation (2012) About CHF: Fact Sheets. Children's Heart Foundation, Lincolnshire. Available at http://www.childrensheartfoundation.org/about-chf/fact-sheets

49. Patterson JT et al. (2012) Tissue-engineered vascular grafts for use in the treatment of congenital heart disease: from the bench to the clinic and back again. Regen Med 7:409–419

50. Webb CL et al. (2002) Collaborative care for adults with congenital heart disease. Circulation 105:2318–2323

51. Connelly MS et al. (1998) Canadian consensus conference on aadult congenital heart disease 1996. Can J Cardiol 14:395–452
52. Chin AJ, Whitehead KK, Watrous RL (2010) Insights after 40 years of the Fontan operation. World J Pediatr Congenit Heart Surg 1:328–343
53. Mayer Jr J et al. (1992) Factors associated with marked reduction in mortality for Fontan operations in patients with single ventricle. J Thorac Cardiovasc Surg 103:444–451.discussion 451-442
54. Adachi I et al. (2005) Fontan operation with a viable and growing conduit using pedicled autologous pericardial roll: serial changes in conduit geometry. J Thorac Cardiovasc Surg 130:1517–1522.e1511
55. Woods RK, Dyamenahalli U, Duncan BW, Rosenthal GL, Lupinetti FM (2003) Comparison of extracardiac Fontan techniques: pedicled pericardial tunnel versus conduit reconstruction. J Thorac Cardiovasc Surg 125:465–471
56. Diodato M, Chedrawy EG (2014) Coronary artery bypass graft surgery: the past, present, and future of myocardial revascularisation. Surg Res Pract 2014:6. https://doi.org/10.1155/2014/726158
57. Lee Y-U et al. (2014) Implantation of inferior vena cava interposition graft in mouse model. J Vis Exp 2014(88):51632. https://doi.org/10.3791/51632
58. Buttafoco L et al. (2006) Physical characterization of vascular grafts cultured in a bioreactor. Biomaterials 27:2380–2389
59. Zhang X et al. (2009) Dynamic culture conditions to generate silk-based tissue-engineered vascular grafts. Biomaterials 30:3213–3223
60. Williams C, Wick TM (2004) Perfusion bioreactor for small diameter tissue-engineered arteries. Tissue Eng 10:930–941
61. Neumann T, Nicholson BS, Sanders JE (2003) Tissue engineering of perfused microvessels. Microvasc Res 66:59–67
62. Hahn MS, McHale MK, Wang E, Schmedlen RH, West JL (2007) Physiologic pulsatile flow bioreactor conditioning of poly (ethylene glycol)-based tissue engineered vascular grafts. Ann Biomed Eng 35:190–200
63. Niklason L et al. (1999) Functional arteries grown in vitro. Science 284:489–493
64. Pashneh-Tala S, MacNeil S, Claeyssens F (2015) The tissue-engineered vascular graft—past, present, and future. Tissue Eng Part B Rev 22:68–100
65. van Hinsbergh VW (2012) Endothelium—role in regulation of coagulation and inflammation. Semin Immunopath 34(1):93–106. https://doi.org/10.1007/s00281-011-0285
66. Brisbois EJ et al. (2015) Reduction in thrombosis and bacterial adhesion with 7 day implantation of S-nitroso-N-acetylpenicillamine (SNAP)-doped Elast-eon E2As catheters in sheep. J Mater Chem B 3:1639–1645
67. Fleser PS et al. (2004) Nitric oxide–releasing biopolymers inhibit thrombus formation in a sheep model of arteriovenous bridge grafts. J Vasc Surg 40:803–811
68. Elliott MB, Gerecht S (2016) Three-dimensional culture of small-diameter vascular grafts. J Mater Chem B 4:3443–3453
69. McFadden T et al. (2013) The delayed addition of human mesenchymal stem cells to pre-formed endothelial cell networks results in functional vascularization of a collagen–glycosaminoglycan scaffold in vivo. Acta Biomater 9:9303–9316
70. Khan OF, Chamberlain MD, Sefton MV (2011) Toward an in vitro vasculature: differentiation of mesenchymal stromal cells within an endothelial cell-seeded modular construct in a microfluidic flow chamber. Tissue Eng A 18:744–756
71. Pedram A, Razandi M, Levin ER (1998) Extracellular signal-regulated protein kinase/Jun kinase cross-talk underlies vascular endothelial cell growth factor-induced endothelial cell proliferation. J Biol Chem 273:26722–26728
72. Eddahibi S et al. (2006) Cross talk between endothelial and smooth muscle cells in pulmonary hypertension critical role for serotonin-induced smooth muscle hyperplasia. Circulation 113:1857–1864

73. Hibino N et al. (2012) Evaluation of the use of an induced puripotent stem cell sheet for the construction of tissue-engineered vascular grafts. J Thorac Cardiovasc Surg 143:696–703
74. Askari F, Solouk A, Shafieian M, Seifalian AM (2017) Stem cells for tissue engineered vascular bypass grafts. Artif Cells Nanomed Biotechnol 45(5):999–1010. https://doi.org/10. 1080/21691401.2016.1198366
75. Bajpai VK, Andreadis ST (2012) Stem cell sources for vascular tissue engineering and regeneration. Tissue Eng Part B Rev 18:405–425
76. Wang Y et al. (2014) Engineering vascular tissue with functional smooth muscle cells derived from human iPS cells and nanofibrous scaffolds. Biomaterials 35:8960–8969
77. Sundaram S et al. (2014) Tissue-engineered vascular grafts created from human induced pluripotent stem cells. Stem Cells Transl Med 3:1535–1543
78. Gui L et al. (2016) Implantable tissue-engineered blood vessels from human induced pluripotent stem cells. Biomaterials 102:120–129
79. Prokhorova TA et al. (2009) Teratoma formation by human embryonic stem cells is site dependent and enhanced by the presence of Matrigel. Stem Cells Dev 18:47–54
80. Sivarapatna A et al. (2015) Arterial specification of endothelial cells derived from human induced pluripotent stem cells in a biomimetic flow bioreactor. Biomaterials 53:621–633
81. Dash BC et al. (2016) Tissue-engineered vascular rings from human iPSC-derived smooth muscle cells. Stem Cell Rep 7:19–28
82. Gui L et al. (2014) Construction of tissue-engineered small-diameter vascular grafts in fibrin scaffolds in 30 days. Tissue Eng A 20:1499–1507
83. Samuel R et al. (2013) Generation of functionally competent and durable engineered blood vessels from human induced pluripotent stem cells. Proc Natl Acad Sci 110:12774–12779
84. Chan XY et al. (2015) Three-dimensional vascular network assembly from diabetic patient-derived induced pluripotent stem cells. Arterioscler Thromb Vasc Biol 35:2677–2685
85. de Carvalho JL et al. (2015) Production of human endothelial cells free from soluble xenogeneic antigens for bioartificial small diameter vascular graft endothelialization. Biomed Res Int 2015:8
86. Lin B et al. (2012) High-purity enrichment of functional cardiovascular cells from human iPS cells. Cardiovasc Res 95:327–335

Adv Biochem Eng Biotechnol (2018) 163: 169–186
DOI: 10.1007/10_2017_25
© Springer International Publishing AG 2017
Published online: 10 November 2017

Targeted Gene Editing in Human Pluripotent Stem Cells Using Site-Specific Nucleases

Sylvia Merkert and Ulrich Martin

Abstract Introduction of induced pluripotent stem cell (iPSC) technology and site-directed nucleases brought a major breakthrough in the development of regenerative therapies and biomedical research. With the advancement of ZFNs, TALENs, and the CRISPR/Cas9 technology, straightforward and precise manipulation of the genome of human pluripotent stem cells (PSC) became possible, allowing relatively easy and fast generation of gene knockouts, integration of transgenes, or even introduction of single nucleotide changes for correction or introduction of disease-specific mutations. We review current applications of site-specific nucleases in human PSCs and focus on trends and challenges for efficient gene editing and improvement of targeting strategies.

S. Merkert and U. Martin (✉)
Department of Cardiothoracic, Transplantation and Vascular Surgery, Leibniz Research
Laboratories for Biotechnology and Artificial Organs (LEBAO), Hannover, Germany

REBIRTH-Cluster of Excellence, German Center for Lung Research (DZL), Gießen, Germany

Hannover Medical School, Hannover, Germany
e-mail: martin.ulrich@mh-hannover.de

Graphical Abstract

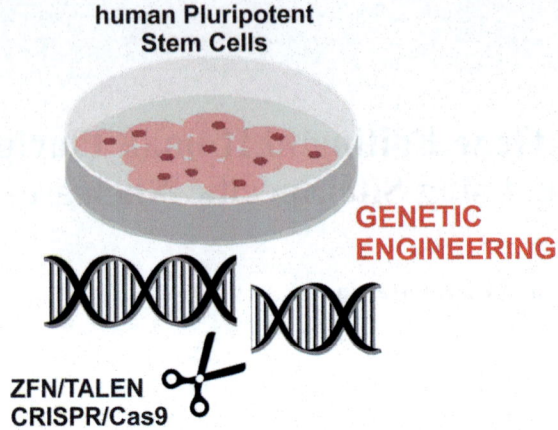

Keywords CRISPR/Cas9, Homologous recombination, NHEJ, TALEN, ZFN

Contents

Abbreviations

2A	Self-cleaving peptide sequence
AAVS1	Adeno-associated virus integration site 1 (safe harbor site)
ABCA1	ATP-binding cassette subfamily A member 1
AKT2	AKT serine/threonine protein kinase 2
ALS	Amyotrophic lateral sclerosis
B2M	Beta-2-microglobulin
CAG	Cytomegalovirus early enhancer element coupled to chicken beta-actin promoter
CAS	CRISPR-associated system
CCR5	C-C motif chemokine receptor 5
CLYBL	Citrate lyase beta-like
CRISPR	Clustered regularly interspaced short palindromic repeats

DNMT3B	DNA methyltransferase 3B
DSB	Double-strand break
EZH2	Enhancer of zeste homolog 2
GATA4	GATA binding protein 4
H3K4/K9	His 3, Lys 4 or Lys 9
HR	Homologous recombination
ICF	Immunodeficiency-centromeric region instability-facial anomalies syndrome
iPSC	Induced pluripotent stem cell
MHC	Myosin heavy chain
NGN3	Neurogenein 3
NHEJ	Nonhomologous end joining
OCT4	Octamer-binding protein 4
PAM	Protospacer adjacent motif
PSC	Pluripotent stem cells
SORT1	Sortilin 1
ssODN	Single-stranded oligonucleotide
TALEN	Transcription activator-like effector nuclease
ZFN	Zinc-finger nuclease

1   Introduction

Human pluripotent stem cells (PSCs), with their unlimited potential for proliferation and differentiation, are the favorite cell source for regenerative therapies and hold great potential for basic research, disease modeling, and drug screening. Until recently, their scope of application was limited because gene transfer and stable transgene expression was very difficult to achieve in these cells. Besides viral transduction, with drawbacks such as laborious vector production, potential site-specific mutagenesis, and frequent transgene gene silencing, only transient transfection and random transgene integration approaches were feasible, again associated with limitations such as unpredictable expression levels and the risk of mutagenesis [1]. Therefore, the development of site-specific and efficient gene editing technologies in combination with more efficient DNA transfection techniques substantially increased the usefulness of human PSCs for experimental research, industrial drug development, and cellular therapies. Targeted genomic modification in human PSCs enables (1) functional knockout of specific genes to investigate gene functions, (2) introduction of reporter and selection genes, (3) introduction of disease-relevant mutations or overexpression of disease-related transgenes, and (4) correction of inherited gene defects in patient-specific induced pluripotent stem cells (iPSCs) or controlled overexpression of therapeutic transgenes.

The frequency of "classical" (i.e., non-nuclease catalyzed) targeted homologous recombination is typically very low (10^{-4} to 10^{-7}), and in human PSCs even lower than in mouse embryonic stem cells (ESCs) or in various immortalized cell lines [2–5].

Hence, the application of positive and negative selection markers to identify cell clones that have undergone these rare events is indispensable. However, the development of protocols for improved plasmid transfection into human PSCs [6], in combination with the introduction of site-specific nucleases for targeted introduction of DNA double-strand breaks (DSBs), led to increased gene targeting efficacy [7–11].

Site-specific nucleases boost the efficiency of homologous recombination events by catalyzing directed DNA DSBs. After the strand is disrupted, the strategy relies on endogenous cellular DNA repair mechanisms and the fact that genomic DSBs can be repaired either by nonhomologous end joining (NHEJ) or by homologous recombination (HR) [12, 13]. This principle is universal and has been successfully applied to human and mouse cells, and to whole organisms such as zebrafish and *Xenopus tropicalis* (reviewed in [14]). The error-prone NHEJ pathway directly ligates the ends of a DNA DSB without the need for a homologous template, which usually leads to accurate repair of the DSB but frequently results in random insertions or deletions. HR requires a homologous repair template (originally facilitated by the sister chromatid), but directed targeting approaches use an exogenous complementary DNA stretch (Fig. 1). Site-specific nucleases act as DNA scissors and are programmable for any desired genomic locus. In consequence, these molecular tools, which include zinc-finger nucleases (ZFNs), transcription activator-like effector nucleases (TALENs), and the clustered regularly

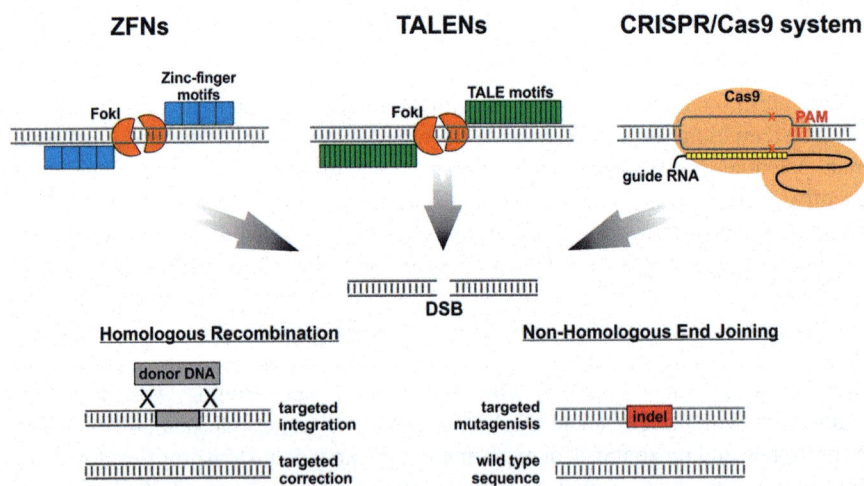

Fig. 1 Site-specific nuclease-induced genome editing. The scheme shows the three types of designer nucleases that have emerged as tools for targeted genome engineering in human PSCs. ZFNs and TALENs consist of DNA-binding domains fused to an unspecific nuclease domain, *Fok*I. In the CRISPR/Cas9 system, a chimeric target-specific RNA guides the Cas9 nuclease to cleave the DNA. Such a nuclease-induced double-strand break (*DSB*) can be repaired either by nonhomologous end joining or by homologous recombination. Nonhomologous end joining can lead to insertions or deletions that disrupt the coding sequence. Targeted gene correction or insertion via homologous recombination can be achieved by the introduction of donor DNA containing homologous sequences of genomic DNA surrounding the DSB

interspaced short palindromic repeats (CRISPR)/CRISPR-associated (Cas) systems, have tremendously increased our ability to manipulate the genome of human PSCs.

2   Current Technologies and Applications

2.1   ZFN, TALEN, and CRISPR/Cas9

In 2002, with the application of ZFNs, the era of "genome editing" began by showing that targeted DNA breaks accelerate the rate of HR [15–18]. A few years later, this advancement was followed by the development of TALENs [19]. Both ZFNs and TALENs carry the same cleaving domain of the bacterial restriction endonuclease FokI, which is active only after dimerization [20] and is guided to the desired genomic loci by a DNA binding domain consisting of an array of zinc-finger or TALE protein motifs (Fig. 1). During this process, one zinc-finger motif recognizes three base pairs, whereas one TALE protein recognizes one base pair. As a ZFN consists of 3–4 zinc-finger motifs and a TALEN consists of 15–20 single repeats, this results in a recognition sequence length of 12 or 20 base pairs, respectively [21]. This sequence length is doubled because a pair of ZFNs or TALENS has to bind the DNA to activate the nuclease domain. The engineering of ZFNs is technically challenging, time consuming, and not suitable for routine laboratory production. Generation of TALENs is less labor intensive once the cloning platform is established. Nevertheless, in terms of assembling the molecular components, the CRISPR/Cas9 system is the easiest because it requires only the design of a 20-bp guide RNA to program the nuclease. In contrast to ZFNs and TALENs, this synthetic single guide RNA directs the Cas9 endonuclease to the complementary 20-nt genomic sequence (also known as protospacer). The sole additional requirement is the presence of a protospacer adjacent motif (PAM) directly downstream of the DNA target sequence, whereby it is not present in the guide RNA sequence (Fig. 1). The Cas9 only binds and cuts the target sequence if both a guide RNA and a PAM are available. If this requirement is fulfilled, the DNA strand is cut three base pairs upstream of the PAM sequence [22].

The most important features of site-specific nucleases are their efficiency and specificity, both relying on the target site selection. High "on-target activity" of the nuclease is strongly desired and must be associated with little or no unspecific off-target activity. In general, the efficiency not only depends on the design of the nuclease, but also to a large extent on the sequence of the genomic target [23]; for certain target sequences, it is difficult or even impossible to design proper site-specific nucleases, as reported for CRISPR/Cas9 and sequences that contain a high proportion of NGG motifs with the potential to form G-quadruplexes [24].

Regarding off-target effects, ZFNs and TALENs can generally be expected to bind fewer unspecific sequences because of the relatively long recognition motifs of

24 or 36 bp, respectively, as compared with the 20-bp recognition sequence of the classical CRISPR/Cas9 system. The development of paired Cas9 "nickases" addressed this lower specificity. Individual nickases only produce single-strand breaks so the targeted activity of two nickases with two recognition sequences is necessary to generate a DSB [25–27]. This drastically reduces the risk of off-site effects because off-target single-strand breaks generated by one isolated nickase are repaired with high fidelity by the "base-excision repair pathway" instead of the error-prone NHEJ [26, 28, 29]. Another possibility for reducing off-target activity in CRISPR/Cas9 targeting is to use "enhanced specificity" SpCas9 variants or a truncated guide RNA of 17–18 nt [30, 31].

2.2 Genetic Engineering in Human Pluripotent Stem Cells

Human PSCs are a key target cell type for the application of accurate genome engineering approaches because of their biomedical importance and unlimited potential for proliferation and differentiation. However, targeted genome editing in human PSCs remained challenging for a long time due to the lack of efficient plasmid transfection protocols, the special growth properties of human PSCs, and their high sensitivity to dissociation and single cell seeding, which is a prerequisite for the derivation of genome-edited cell clones. For several years, lentiviral transduction was the method of choice for introduction of genome-integrated transgenes into PSCs. However, the expression of lentiviral vectors is unpredictable because of transgene silencing and random integration into the genome of host cells. Traditional gene targeting approaches via HR require relatively large homologous DNA stretches of several hundred to several thousand base pairs and the utilization of genetic selection markers, and the efficiencies are low [5, 32]. These hurdles were ultimately overcome by the application of advanced electroporation protocols [33], application of the ROCK inhibitor Y-27632 to support single cell survival of human PSCs [34], and development of site-specific nucleases. Since the first successful application of ZFNs in human PSCs [16, 17, 35], numerous studies have been conducted using ZFNs, TALENs, or the CRISPR/Cas9 system for targeted genome engineering. These tools have been applied for diverse gene correction approaches, integration of specific mutations, overexpression of transgenes, introduction of reporter or selection genes, and generation of gene knockouts.

However, despite this progress, it must be emphasized that gene editing in human PSCs is still no trivial approach and that some crucial aspects have to be considered. First, the most suitable nuclease for the specific approach of interest must be chosen. For many loci, ZFNs or TALENs are already available and ready to use; but the CRISPR/Cas9 system also supports the easy and fast design of guide RNAs for any desired locus. Certainly, the most important point is to achieve sufficient nuclease expression in human PSCs. This is especially crucial in case of HR-based strategies and, in particular, when targeting of both alleles is necessary (e.g., for knockout approaches). Hence, experience with different human PSCs culture systems (e.g., feeder-based or feeder-free cultivation) and efficient transfection methods (e.g., lipofection or electroporation-based nucleofection) are advantageous for achieving successful gene targeting.

The overall targeting strategy is determined by the overall goal of the project, which may be a "simple" loss-of-function knockout, introduction of a transgene under the control of an exogenous promoter, knockin of a transgene under the control of an endogenous promoter, or introduction of seamless single nucleotide changes. For selection-free approaches, which do not allow transgene reporter-based enrichment of targeted cell clones, the targeting efficacy must be high to facilitate the likelihood of identifying targeted clones by random screening. To achieve this goal, preselection of nuclease-expressing cells is needed, for example by co-transfection with an enhanced green fluorescent protein (eGFP) expression plasmid or an expression vector with a reporter gene coupled to the nuclease via a 2A site. Integration of selection markers can be very useful for identification of positive targeted cells but may require a second round of single cell cloning because of the need to remove the targeting cassette, if a transgene-free or selection marker-free result is needed.

2.2.1 Gene Disruption/Generation of Gene Knockouts

The most common and straightforward targeted gene editing approach in human PSCs is the generation of gene knockouts. The strategy allows investigation of the function of an individual gene in human PSCs and their differentiated derivatives. This goal can be achieved simply by introducing a locus-specific DSB followed by NHEJ, assuming that repair of the introduced DSB can lead to random base pair insertions or deletions that consequently code for nonsense mutations or premature stop codons. Therefore, such an approach can either lead to the expression of nonfunctional mRNA or completely prevent transcription of the mRNA.

Despite choosing the appropriate nuclease, there are different general strategies for achieving successful loss of gene function. There may be different splice variants of a target gene, therefore a strategy that covers all these variants must be designed. If the role of splice variants is to be analyzed, these have to be targeted specifically. Frequently, important functional domains of a protein are targeted, or the start codon and downstream sequences are eliminated. In a less-specific but straightforward approach utilizing the NHEJ pathway, DSBs are placed in the relevant areas to induce deletions or insertions that, for instance, result in loss of the start codon. Introducing two DSBs at distinct sites is another strategy for removing larger stretches of the target gene or even the depletion of large genomic regions in the range of several hundred kilobase pairs [36]. For practical reasons, one can also use HR to generate knockouts via the integration of a selection cassette to enrich for correctly targeted clones.

Aiming at in vitro disease modeling or general understanding of gene function, the above strategy was successfully applied for the inactivation of several genes including AKT2 (AKT serine/threonine-protein kinase 2; insulin resistance),

SORT1 (Sortilin 1; modulate blood glucose and cholesterol levels), DNMT3B (DNA methyltransferase 3B; disease model for immunodeficiency-centromeric region instability-facial anomalies (ICF) syndrome), and ABCA1 (ATP-binding cassette subfamily A member 1; model for Tangier disease) [37–41]. It is also very common to use an inducible Cas9 system to generate single or even multiple gene knockouts. Therefore, a doxycycline-inducible expression cassette for the Cas9 nuclease was introduced into the AAVS1 (adeno-associated virus integration site 1) safe harbor site, resulting in universal cell lines that could be further used for efficient generation of knockout cell lines [42]. By simple expression or delivery of single guide RNAs and simultaneous application of doxycycline to trigger Cas9 expression, gene knockouts for NGN3 (neurogenein 3), GATA4 (GATA binding protein 4), EZH2 (enhancer of zeste homolog 2), OCT4 (octamer-binding protein 4), B2M (beta-2-microglobulin), and Brachyury were achieved, enabling analysis of embryogenesis, myocardial differentiation, hematopoietic differentiation, pluripotency, immunogenicity, and mesendoderm formation, respectively [42–44].

2.2.2 Gene Insertion/Knockin Approaches

For the integration of transgenes into specific genomic loci, donor DNA has to be constructed to serve as a template for the HR pathway. Donor DNA can be provided either as plasmid DNA or as single-stranded oligonucleotides (ssODN). In combination with site-directed nucleases, the donor template carries the transgene flanked by short homologous sequences of up to 500 bp on plasmid DNA or 50 bp on ssODNs. The homology-directed recombination can be applied for the integration of numerous different transgenes, including fluorescence markers, resistance genes, suicide genes, recombinase genes, therapeutic transgenes, protein tags, and recombination sites. Human PSC reporter lines are useful tools for the real-time tracking of gene expression in pluripotent cells, especially during differentiation processes. They enable the optimization of differentiation protocols, purification of cell populations, or monitoring of cell survival, distribution, and integration in vivo. Application of an HES3-NKX2.5$^{w/GFP}$ reporter cell line enabled optimization of cardiomyogenic differentiation in scalable suspension cultures [45]. Further reporter cell lines have been generated for OCT4 (pluripotency marker), LGR5 (intestinal stem cells), and MYH5 or PAX7 (myogenic linages) [19, 46–48]. In addition to reporter cell lines employing transgene integration into endogenous gene loci, HR can also be used for gene insertions into safe harbor sites, enabling robust transgene expression, for example, for overexpression of therapeutic transgenes [49–51] aiming at ex vivo gene therapy (see also Sect. 3).

2.2.3 Gene Correction and Point Mutagenesis

The alteration of just a few nucleotides without any footprints is applied for precise gene correction in disease-specific iPSCs or for a reverse approach, such as the introduction of disease-specific or disease-associated modifications into a wild-type genetic background. To achieve this goal, ssODNs or selection genes (with subsequent removal of the selection cassette) are the methods of choice. The application of ssODNs was successfully used for the correction of disease-related mutations in iPSC lines from patients with X-linked chronic granulomatous disease, amyotrophic lateral sclerosis (ALS), retinitis pigmentosa, and sickle-cell disease [52–55]. In most cases, the application of appropriate expression plasmids for Cas9 nucleases or nickases enabled preselection of successfully transfected cells, thereby substantially facilitating the identification of correctly targeted cells. Similar strategies were applied for the integration of specific mutations to model diseases such as ALS, the mitochondrial cardiomyopathy of Barth syndrome, or Parkinson's disease [56–58]. Besides the use of ssODNs, the co-transfection of a donor plasmid for selection and subsequent removal of the selection cassette can also be applied for precise base pair correction. This can be achieved by using either the Cre/LoxP or FLP/FRT recombinase system (notably, both systems leave a "genomic scar" of 34 bp) [59–61] or the piggyBac transposon flanking the selectable marker. The latter approach restores its original insertion site after remobilization of the vector, thus enabling seamless excision of the introduced selection transgenes [62–64].

2.3 Improvement and Extended Applications of the CRISPR/Cas9 System

For targeting approaches in human PSCs, all three site-directed nuclease types described above are suitable. Their targeting efficiencies are in principle quite similar and the targeting success seems to depend more on the accessibility of the genomic locus of interest and the researcher's cell culture skills [16, 19, 37, 38]. Generation of ZFNs and TALENs, however, requires more experience in cloning efforts. This is a key reason for the broad application of the more easily applicable CRISPR/Cas9 system for genome editing. In addition, the CRISPR system is increasingly attractive because of continuous advancement of the technology and the development of a broader spectrum of applications.

The CRISPR/Cas system originates from the "bacterial immune system," protecting prokaryotic cells against foreign DNA, in particular invading bacteriophages. The power and attractiveness of the technique for biological research lies in the binding specificity of the approach, which is simply determined by standard RNA/DNA base pairing.

However, the CRISPR/Cas9 system is more than just a site-directed nuclease. Its specific recognition ability can be employed for many research applications beyond targeted mutation or transgene integration. For instance, a catalytically inactive form of Cas9 ("dead" Cas9 or dCas9) can bind to specific DNA loci without cutting [65]. An interesting application of this approach is the fusion of a cytidine deaminase to dCas9, thereby enabling the modification of a single base [66, 67] without utilizing the NHEJ or HR pathways. Other applications include fusion of transcriptional repressor or activator domains to dCas9, which enables gene silencing or targeted activation of gene expression, respectively [68–70]. Thus, CRISPR-mediated transcriptional repression also enables targeting of noncoding regulatory sequences such as distant enhancers and locus-control regions, providing novel possibilities for targeted gene regulation beyond the RNA interference (RNAi) system, which directly targets coding and noncoding RNAs. Notably, such CRISPR interference has already been applied in human iPSCs to enable inducible, efficient gene knockdown at the pluripotent state and in differentiated cell derivatives [71]. Hence, fusion with activators and repressors also allows loss-of-function and gain-of-function screens in human cells [68]. Moreover, it enables CRISPR interference-based screening for identification of functional long-noncoding RNA loci in human iPSCs [72]. Alternatively, dCas9 can be fused with the histone demethylase enzyme LDS1, allowing demethylation of His H3, Lys 4 or Lys 9 (H3K4/K9) at enhancer sequences and, thus, gene repression via CRISPR-mediated epigenetic regulation [73]. In another approach, visualization of genomic loci in vivo became possible through combination of dCas9 with eGFP [74]. This range of attractive CRISPR-based tools can lead to a multitude of exciting novel applications in human PSCs, such as identifying pluripotency regulators and new developmental pathways, understanding cellular processes, and providing new insights into the pathogenesis of human disease.

3   Trends and Strategies for Improving Genome Targeting

Genome editing via site-directed nucleases, DNA-modifying enzymes, and expression modulators offers an enormous field of applications for human PSCs. However, targeting in human PSCs is still challenging because of the specific culture requirements of human PSCs compared with more easily handled cell systems such as mouse PSCs and human cell lines like HEK293T or Hela cells. Because the targeting efficacies of such cell lines are relative high, investigators are often surprised when they face challenges in generating genetically engineered human PSC lines. One should not underestimate the crucial requirement to optimize culture conditions, transfection protocols, and cell dissociation for single cell cloning, which are all of utmost importance for the technology to work. Moreover, extensive experience in assessment of the quality of human PSC cultures is required. In laboratory practice, it is often unrealistic to aim for published targeting efficiencies; leading investigators in the field are usually very experienced and often work with well-established cell lines and standardized and adapted culture systems.

Investigators that are new to the field are therefore advised to use established PSC lines that show good transfection efficiencies and have already demonstrated tolerance for single cell cloning. For instance, the application of stable human PSC lines readily carrying a doxycycline-inducible Cas9 expression cassette in the

AAVS1 locus should substantially increase the chances of successful targeting [42]. Targeting efficiencies always depend on the efficiency of the nuclease and the accessibility of the locus of interest. In general, targeting approaches employing selection strategies are more successful and less laborious than selection-free approaches requiring PCR-based screening of hundreds of clones. Possible selection strategies include (1) preselection of nuclease-expressing cells using co-expression of fluorescent markers or antibiotic resistance cassettes, and (2) positive and negative selection for targeted cell clones by integration of additional reporters into the donor targeting cassette. The latter comprises resistance genes (e.g., neomycin resistance) or suicide genes (e.g., thymidine kinase), which can be removed after successful targeting if required for the intended application of the targeted cell clones (see Sect. 2.2.3).

Another possibility for increasing targeting efficiency is application of small molecules acting on proteins that are involved in the cellular repair mechanisms. If the targeting strategy aims to utilize HR instead of NHEJ, it is important to take into account that NHEJ-based repair of DSBs is active during the whole cell cycle but HR occurs only during S and G2 phases. Hence, specific inhibition of NHEJ or synchronizing and shifting the target cells into S and G2 phase should lead to a relative increase in DSBs repaired through HR. It was shown that inhibition or suppression of key enzymes in the NHEJ pathway by inhibitors (Scr7) or short hairpin RNAs is a viable method for the enhancement of HR-mediated genome targeting using Cas9 in mammalian cells [75, 76]. Another study identified several small molecules (e.g., L755507, Brefeldin A) that enhance CRISPR-mediated precise genome editing in mouse ESCs by interacting with both repair pathways [56].

In addition to high on-target efficiencies, the off-target activity of the applied nuclease is also crucial because it can lead to unwanted mutagenesis, which is a particular problem for cellular products for clinical application. For experimental use of edited cell lines, including disease modeling and drug development, this issue is less critical if (1) proper controls are applied and (2) mutations of the computationally predicted "top 10 off-target sites" of the respective nuclease can be excluded by screening. An overview of methods for measuring off-target effects in the genome, such as chromatin immunoprecipitation coupled with deep sequencing, systematic evolution of ligands by exponential amplification, and whole genome/exome sequencing, is provided in recent reviews from Koo et al. and Lee et al. [77, 78]. However, the limited sensitivity of the available screening assays (at best 0.1% of all events) emphasize that it is indispensable to intensify the development of more sensitive approaches for detection of off-target effects and to carefully investigate functional deficits and the potential for tumor formation in clinically applicable gene-edited PSC derivatives [23].

In addition to off-target nuclease activity, the application of transfection plasmids for nuclease expression and donor DNA delivery also bears the risk of additional random integrations and, thus, of dysregulating transgene expression or insertional mutagenesis. Transfection of recombinant Cas9/TALEN proteins is an alternative approach but there are typically considerable batch-to-batch variations in the quality of recombinant proteins, and the transfection efficacy is typically very low [79, 80]. Another potentially more reproducible approach is the transfection of synthetic mRNA encoding for Cas9/TALEN [81, 82].

In general, random integration of transgenes cannot be recommended because of unpredictable expression levels and high risk of insertional mutagenesis leading to altered cell function or tumorigenesis. Hence, safe harbor sites are of particular interest for the introduction of reporter genes or therapeutic transgenes. Safe harbor sites such the AAVS1 locus, the C-C motif chemokine receptor 5 (CCR5) locus, the ROSA26 locus, or the citrate lyase beta-like (CLYBL) locus are safe for the introduction of transgenes, with no reported phenotypic effects on treated cells. Moreover, such loci support robust transgene expression, not only in undifferentiated cells but also in their differentiated derivatives [43, 83–86]. They enable overexpression of therapeutic transgenes for functional correction of genetic diseases or even tissue- or cell type-specific transgene expression if a well-performing promoter fragment is available, which also facilitate isolation of specific cell populations during differentiation approaches. In our hands, safe harbor sites proved very useful for the constitutive labeling of cells, enabling their monitoring in engineered tissue or in vivo in animal models (unpublished). Here, robust transgene expression allows monitoring of graft survival, cell distribution, and functional integration. Even differentiation processes or proliferation can be followed if suitable reporters are used. For instance, we inserted fluorescent proteins such as eGFP and RedStar under control of the CAG promoter into the AAVS1 locus of human iPSC lines via TALENs [84]. Stable transgenic cell lines could be established via cell sorting. In these lines, fluorescent protein expression is stable in undifferentiated iPSCs and in differentiated derivatives (Fig. 2), including cardiomyocytes, endothelial cells, epithelial cells, and macrophages [84, 87]. As an example, we successfully applied a RedStar iPSC line [88] for differentiation into cardiomyocytes and an eGFP-labeled iPSC line for differentiation into endothelial cells, and applied these populations for the generation of bioartificial cardiac tissue [89]. The two colors allowed exploration of the distribution, survival, and structural integration of both cell populations in the constructs and enabled optimization of the tissue engineering process (unpublished). Another example is the integration of a cardiomyocyte selection cassette into one allele of the AAVS1 locus containing a neomycin resistance gene under control of the cardiac α-myosin heavy chain (αMHC) promoter fragment. In this case, AAVS1-specific TALENs, a donor vector with the neomycin resistance gene, and a PGK promotor-driven hygromycin resistance for selection during clone establishment were applied. It is noteworthy that the specificity of the αMHC promoter was maintained in the AAVS1 locus, allowing antibiotic selection of almost pure cardiomyocytes from human PSCs (unpublished).

Like many other research groups, we employ the AAVS1 locus, which is located on chromosome 19 in the first intron of the PPP1R12C gene. The locus is described as open chromatin, with insulator activity ascribed to a DNase hypersensitivity region [90]. Nevertheless, there are reports of partial silencing of transgene expression in this locus, which might be promoter or transgene dependent [91]. We

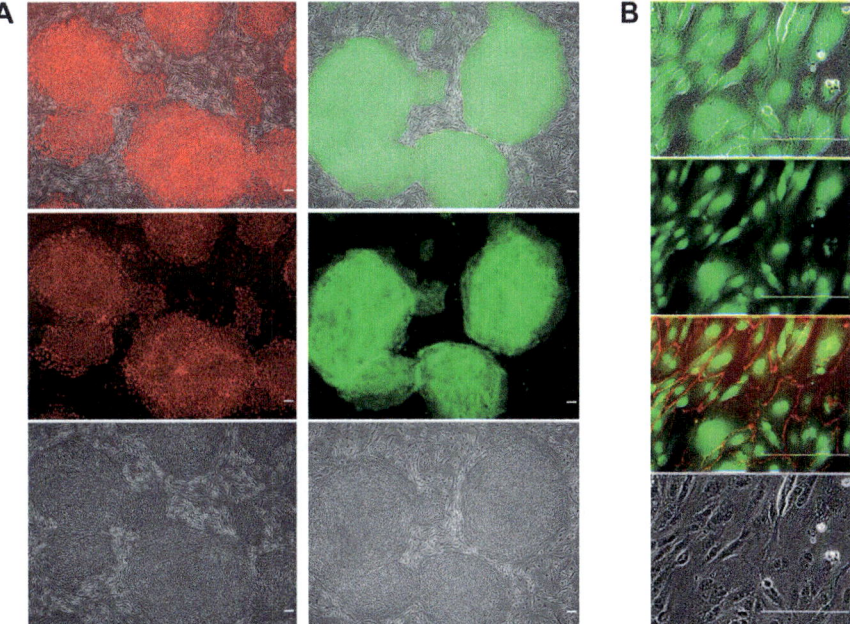

Fig. 2 Stable transgene expression from the AAVS1 locus in undifferentiated iPSC clones and in iPSC-derived endothelial cells. (**a**) Microscopy images of transgenic human CBiPS2-RSC8 expressing nuclear RedStar (*left column*) and human CBiPS2-eGFPC18 expressing eGFP (*right column*). Phase contrast with overlay of the respective fluorescence (*upper row*), fluorescence only (*middle row*), and phase contrast only (*lower row*). *Scale bars* represent 100 μm. (**b**) Microscopy images of human CBiPS2-eGFPC18-derived endothelial cells. *Top-down*: Phase contrast with overlay of eGFP fluorescence, eGFP fluorescence only, eGFP fluorescence and VE-Cadherin staining (*red*), and phase contrast only. *Scale bars* represent 100 μm

observed that the CAG promoter in the AAVS1 locus always operates reliably during differentiation. Similarly, the cardiomyocyte-specific αMHC promoter works properly in our established transgenic cell lines. However, we also found single clones without selection specificity, which might be the result of epigenetic changes.

Another crucial aspect that should be mentioned is that the culture characteristics and differentiation behavior of targeted transgenic cell clones can be completely different to those of the nontargeted parental cell line. In most cases, however, this is not caused by the integrated transgene, but by culture adaptation of individual clones during the transfection and single cell cloning processes. The resulting transgenic cell lines represent subclones, which might carry de novo mutations or epigenetic changes that allow better survival during the cloning process or increased proliferation under the applied conditions. For example, we found that human iPSC clones, highly efficient in cardiomyocyte differentiation, can lose their cardiac differentiation ability after integration of a fluorescent reporter in the AAVS1 locus (unpublished). Exclusion of additional random integration of large transgene fragments in the genome as underlying reason for loss of differentiation ability is easy to perform via PCR and southern blot analysis; however, small genetic changes cannot be detected in this way and whole genome sequencing of each clone is still relatively expensive and the necessary bioinformatics analyses are very laborious. Hence, the establishment and functional testing of three to five different clones is always required, because single cell cloning may select for a specific genomic context and the clonal differences may have a severe impact on culture behavior and differentiation capacity.

4 Conclusion

The introduction of site-specific nucleases and their application in human PSCs was an enormous stimulus to the development of the entire field of stem cell research. Targeted gene editing and gene-edited PSCs and their derivatives are exciting tools in basic research, disease modeling, and drug screening endeavors. Precise genome editing is expected to push forward the therapeutic application of cellular therapeutics for ex vivo gene therapy, including application of gene-corrected patient-specific iPSCs, in particular addressing as yet incurable and fatal diseases. However, despite the enormous technical progress in the field, gene targeting in human PSCs is still technically challenging. For this reason, investigators require profound knowledge of human PSC culture and targeting strategies, procedures, and efforts to achieve successful genome editing.

Acknowledgments We thank R. Zweigerdt for contributing to fruitful discussions and critical reading of the manuscript, and R. Olmer for providing microscopy images of endothelial cells.

References

1. Hacein-Bey-Abina S et al. (2008) Insertional oncogenesis in 4 patients after retrovirus-mediated gene therapy of SCID-X1. J Clin Invest 118(9):3132–3142
2. Vasquez KM et al. (2001) Manipulating the mammalian genome by homologous recombination. Proc Natl Acad Sci U S A 98(15):8403–8410
3. Reid LH et al. (1991) Cotransformation and gene targeting in mouse embryonic stem cells. Mol Cell Biol 11(5):2769–2777
4. Doetschman T, Maeda N, Smithies O (1988) Targeted mutation of the Hprt gene in mouse embryonic stem cells. Proc Natl Acad Sci U S A 85(22):8583–8587
5. Zwaka TP, Thomson JA (2003) Homologous recombination in human embryonic stem cells. Nat Biotechnol 21(3):319–321
6. Schwanke K et al. (2014) Fast and efficient multitransgenic modification of human pluripotent stem cells. Hum Gene Ther Methods 25(2):136–153
7. Johnson RD, Jasin M (2001) Double-strand-break-induced homologous recombination in mammalian cells. Biochem Soc Trans 29(Pt 2):196–201

8. Rouet P, Smih F, Jasin M (1994) Expression of a site-specific endonuclease stimulates homologous recombination in mammalian cells. Proc Natl Acad Sci U S A 91(13):6064–6068

9. Cho SW et al. (2013) Targeted genome engineering in human cells with the Cas9 RNA-guided endonuclease. Nat Biotechnol 31(3):230–232

10. Durai S et al. (2005) Zinc finger nucleases: custom-designed molecular scissors for genome engineering of plant and mammalian cells. Nucleic Acids Res 33(18):5978–5990

11. Porteus MH, Carroll D (2005) Gene targeting using zinc finger nucleases. Nat Biotechnol 23(8):967–973

12. Shrivastav M, De Haro LP, Nickoloff JA (2008) Regulation of DNA double-strand break repair pathway choice. Cell Res 18(1):134–147

13. Jeggo PA (1998) DNA breakage and repair. Adv Genet 38:185–218

14. Segal DJ, Meckler JF (2013) Genome engineering at the dawn of the golden age. Annu Rev Genomics Hum Genet 14:135–158

15. Urnov FD et al. (2005) Highly efficient endogenous human gene correction using designed zinc-finger nucleases. Nature 435(7042):646–651

16. Hockemeyer D et al. (2009) Efficient targeting of expressed and silent genes in human ESCs and iPSCs using zinc-finger nucleases. Nat Biotechnol 27(9):851–857

17. Zou J et al. (2009) Gene targeting of a disease-related gene in human induced pluripotent stem and embryonic stem cells. Cell Stem Cell 5(1):97–110

18. Bibikova M et al. (2002) Targeted chromosomal cleavage and mutagenesis in drosophila using zinc-finger nucleases. Genetics 161(3):1169–1175

19. Hockemeyer D et al. (2011) Genetic engineering of human pluripotent cells using TALE nucleases. Nat Biotechnol 29(8):731–734

20. Smith J et al. (2000) Requirements for double-strand cleavage by chimeric restriction enzymes with zinc finger DNA-recognition domains. Nucleic Acids Res 28(17):3361–3369

21. Gaj T, Gersbach CA, Barbas 3rd CF (2013) ZFN, TALEN, and CRISPR/Cas-based methods for genome engineering. Trends Biotechnol 31(7):397–405

22. Mali P et al. (2013) RNA-guided human genome engineering via Cas9. Science 339(6121):823–826

23. CRISPR catch-up (2017). Nat Biotechnol 35(5):389

24. Malina A et al. (2015) PAM multiplicity marks genomic target sites as inhibitory to CRISPR-Cas9 editing. Nat Commun 6:10124

25. Merkle FT et al. (2015) Efficient CRISPR-Cas9-mediated generation of Knockin human pluripotent stem cells lacking undesired mutations at the targeted locus. Cell Rep 11(6):875–883

26. Shen B et al. (2014) Efficient genome modification by CRISPR-Cas9 nickase with minimal off-target effects. Nat Methods 11(4):399–402

27. Chen Y et al. (2015) Engineering human stem cell lines with inducible gene knockout using CRISPR/Cas9. Cell Stem Cell 17(2):233–244

28. Mali P et al. (2013) CAS9 transcriptional activators for target specificity screening and paired nickases for cooperative genome engineering. Nat Biotechnol 31(9):833–838

29. Ran FA et al. (2013) Double nicking by RNA-guided CRISPR Cas9 for enhanced genome editing specificity. Cell 154(6):1380–1389

30. Slaymaker IM et al. (2016) Rationally engineered Cas9 nucleases with improved specificity. Science 351(6268):84–88

31. Fu Y et al. (2014) Improving CRISPR-Cas nuclease specificity using truncated guide RNAs. Nat Biotechnol 32(3):279–284

32. Di Domenico AI et al. (2008) Sequential genetic modification of the hprt locus in human ESCs combining gene targeting and recombinase-mediated cassette exchange. Cloning Stem Cells 10(2):217–230

33. Xu XQ et al. (2008) Highly enriched cardiomyocytes from human embryonic stem cells. Cytotherapy 10(4):376–389

34. Watanabe K et al. (2007) A ROCK inhibitor permits survival of dissociated human embryonic stem cells. Nat Biotechnol 25(6):681–686

35. Lombardo A et al. (2007) Gene editing in human stem cells using zinc finger nucleases and integrase-defective lentiviral vector delivery. Nat Biotechnol 25(11):1298–1306
36. Byrne SM et al. (2015) Multi-kilobase homozygous targeted gene replacement in human induced pluripotent stem cells. Nucleic Acids Res 43(3):e21
37. Ding Q et al. (2013) A TALEN genome-editing system for generating human stem cell-based disease models. Cell Stem Cell 12(2):238–251
38. Ding Q et al. (2013) Enhanced efficiency of human pluripotent stem cell genome editing through replacing TALENs with CRISPRs. Cell Stem Cell 12(4):393–394
39. Horii T et al. (2013) Generation of an ICF syndrome model by efficient genome editing of human induced pluripotent stem cells using the CRISPR system. Int J Mol Sci 14(10): 19774–19781
40. Liao J et al. (2015) Targeted disruption of DNMT1, DNMT3A and DNMT3B in human embryonic stem cells. Nat Genet 47(5):469–478
41. Gupta RM et al. (2016) Genome-edited human pluripotent stem cell-derived macrophages as a model of reverse cholesterol transport-brief report. Arterioscler Thromb Vasc Biol 36(1): 15–18
42. Gonzalez F et al. (2014) An iCRISPR platform for rapid, multiplexable, and inducible genome editing in human pluripotent stem cells. Cell Stem Cell 15(2):215–226
43. Bertero A et al. (2016) Optimized inducible shRNA and CRISPR/Cas9 platforms for in vitro studies of human development using hPSCs. Development 143(23):4405–4418
44. Kotini AG et al. (2015) Functional analysis of a chromosomal deletion associated with myelo-dysplastic syndromes using isogenic human induced pluripotent stem cells. Nat Biotechnol 33(6):646–655
45. Kempf H et al. (2014) Controlling expansion and cardiomyogenic differentiation of human pluripotent stem cells in scalable suspension culture. Stem Cell Rep 3(6):1132–1146
46. Forster R et al. (2014) Human intestinal tissue with adult stem cell properties derived from pluripotent stem cells. Stem Cell Rep 2(6):838–852
47. Wu J et al. (2016) Generation and characterization of a MYF5 reporter human iPS cell line using CRISPR/Cas9 mediated homologous recombination. Sci Rep 6:18759
48. Wu J et al. (2016) Generation and validation of PAX7 reporter lines from human iPS cells using CRISPR/Cas9 technology. Stem Cell Res 16(2):220–228
49. Zou J et al. (2011) Oxidase-deficient neutrophils from X-linked chronic granulomatous disease iPS cells: functional correction by zinc finger nuclease-mediated safe harbor targeting. Blood 117(21):5561–5572
50. Chang CJ, Bouhassira EE (2012) Zinc-finger nuclease-mediated correction of alpha-thalassemia in iPS cells. Blood 120(19):3906–3914
51. Dreyer AK et al. (2015) TALEN-mediated functional correction of X-linked chronic granulo-matous disease in patient-derived induced pluripotent stem cells. Biomaterials 69:191–200
52. Flynn R et al. (2015) CRISPR-mediated genotypic and phenotypic correction of a chronic granulomatous disease mutation in human iPS cells. Exp Hematol 43(10):838–848 e3
53. Bhinge A et al. (2017) Genetic correction of SOD1 mutant iPSCs reveals ERK and JNK acti-vated AP1 as a driver of neurodegeneration in amyotrophic lateral sclerosis. Stem Cell Rep 8(4):856–869
54. Bassuk AG et al. (2016) Precision medicine: genetic repair of retinitis Pigmentosa in patient-derived stem cells. Sci Rep 6:19969
55. Li C et al. (2016) Novel HDAd/EBV reprogramming vector and highly efficient ad/CRISPR-Cas sickle cell disease gene correction. Sci Rep 6:30422
56. Yu C et al. (2015) Small molecules enhance CRISPR genome editing in pluripotent stem cells. Cell Stem Cell 16(2):142–147
57. Wang G et al. (2014) Modeling the mitochondrial cardiomyopathy of Barth syndrome with induced pluripotent stem cell and heart-on-chip technologies. Nat Med 20(6):616–623
58. Soldner F et al. (2011) Generation of isogenic pluripotent stem cells differing exclusively at two early onset Parkinson point mutations. Cell 146(2):318–331

59. Huang X et al. (2015) Production of gene-corrected adult Beta globin protein in human erythrocytes differentiated from patient iPSCs after genome editing of the sickle point mutation. Stem Cells 33(5):1470–1479
60. Sebastiano V et al. (2011) In situ genetic correction of the sickle cell anemia mutation in human induced pluripotent stem cells using engineered zinc finger nucleases. Stem Cells 29(11):1717–1726
61. Ma N et al. (2013) Transcription activator-like effector nuclease (TALEN)-mediated gene correction in integration-free beta-thalassemia induced pluripotent stem cells. J Biol Chem 288(48): 34671–34679
62. Xie F et al. (2014) Seamless gene correction of beta-thalassemia mutations in patient-specific iPSCs using CRISPR/Cas9 and piggyBac. Genome Res 24(9):1526–1533
63. Firth AL et al. (2015) Functional gene correction for cystic fibrosis in lung epithelial cells generated from patient iPSCs. Cell Rep 12(9):1385–1390
64. Ye L et al. (2014) Seamless modification of wild-type induced pluripotent stem cells to the natural CCR5Delta32 mutation confers resistance to HIV infection. Proc Natl Acad Sci U S A 111(26):9591–9596
65. Qi LS et al. (2013) Repurposing CRISPR as an RNA-guided platform for sequence-specific control of gene expression. Cell 152(5):1173–1183
66. Kim YB et al. (2017) Increasing the genome-targeting scope and precision of base editing with engineered Cas9-cytidine deaminase fusions. Nat Biotechnol 35(4):371–376
67. Komor AC et al. (2016) Programmable editing of a target base in genomic DNA without double-stranded DNA cleavage. Nature 533(7603):420–424
68. Gilbert LA et al. (2014) Genome-scale CRISPR-mediated control of gene repression and activation. Cell 159(3):647–661
69. Gilbert LA et al. (2013) CRISPR-mediated modular RNA-guided regulation of transcription in eukaryotes. Cell 154(2):442–451
70. Maeder ML et al. (2013) CRISPR RNA-guided activation of endogenous human genes. Nat Methods 10(10):977–979
71. Mandegar MA et al. (2016) CRISPR interference efficiently induces specific and reversible gene silencing in human iPSCs. Cell Stem Cell 18(4):541–553
72. Liu SJ et al. (2017) CRISPRi-based genome-scale identification of functional long noncoding RNA loci in human cells. Science 355(6320):aah7111
73. Kearns NA et al. (2015) Functional annotation of native enhancers with a Cas9-histone demethylase fusion. Nat Methods 12(5):401–403
74. Chen B et al. (2013) Dynamic imaging of genomic loci in living human cells by an optimized CRISPR/Cas system. Cell 155(7):1479–1491
75. Maruyama T et al. (2015) Increasing the efficiency of precise genome editing with CRISPR-Cas9 by inhibition of nonhomologous end joining. Nat Biotechnol 33(5):538–542
76. Chu VT et al. (2015) Increasing the efficiency of homology-directed repair for CRISPR-Cas9-induced precise gene editing in mammalian cells. Nat Biotechnol 33(5):543–548
77. Lee CM et al. (2016) Nuclease target site selection for maximizing on-target activity and minimizing off-target effects in genome editing. Mol Ther 24(3):475–487
78. Koo T, Lee J, Kim JS (2015) Measuring and reducing off-target activities of programmable nucleases including CRISPR-Cas9. Mol Cells 38(6):475–481
79. Ru R et al. (2013) Targeted genome engineering in human induced pluripotent stem cells by penetrating TALENs. Cell Regener J 2(1):5
80. Kim S et al. (2014) Highly efficient RNA-guided genome editing in human cells via delivery of purified Cas9 ribonucleoproteins. Genome Res 24(6):1012–1019
81. Yin H et al. (2016) Therapeutic genome editing by combined viral and non-viral delivery of CRISPR system components in vivo. Nat Biotechnol 34(3):328–333
82. Mock U et al. (2015) mRNA transfection of a novel TAL effector nuclease (TALEN) facilitates efficient knockout of HIV co-receptor CCR5. Nucleic Acids Res 43(11):5560–5571

83. Yao Y et al. (2012) Generation of CD34+ cells from CCR5-disrupted human embryonic and induced pluripotent stem cells. Hum Gene Ther 23(2):238–242
84. Merkert S et al. (2014) Efficient designer nuclease-based homologous recombination enables direct PCR screening for footprintless targeted human pluripotent stem cells. Stem Cell Rep 2(1):107–118
85. Smith C et al. (2015) Efficient and allele-specific genome editing of disease loci in human iPSCs. Mol Ther 23(3):570–577
86. Cerbini T et al. (2015) Transcription activator-like effector nuclease (TALEN)-mediated CLYBL targeting enables enhanced transgene expression and one-step generation of dual reporter human induced pluripotent stem cell (iPSC) and neural stem cell (NSC) lines. PLoS One 10(1):e0116032
87. Ackermann M et al. (2017) Ex vivo generation of genetically modified macrophages from human induced pluripotent stem cells. Transfus Med Hemother 44(3):135–142
88. Knop M et al. (2002) Improved version of the red fluorescent protein (drFP583/DsRed/RFP). BioTechniques 33(3):592. 594, 596-8 passim
89. Kensah G et al. (2013) Murine and human pluripotent stem cell-derived cardiac bodies form contractile myocardial tissue in vitro. Eur Heart J 34(15):1134–1146
90. Ogata T, Kozuka T, Kanda T (2003) Identification of an insulator in AAVS1, a preferred region for integration of adeno-associated virus DNA. J Virol 77(16):9000–9007
91. Ordovas L et al. (2015) Efficient recombinase-mediated cassette exchange in hPSCs to study the hepatocyte lineage reveals AAVS1 locus-mediated transgene inhibition. Stem Cell Rep 5(5):918–931

Adv Biochem Eng Biotechnol (2018) 163: 187–206
DOI: 10.1007/10_2017_22
© Springer International Publishing AG 2017
Published online: 26 October 2017

Acquired Genetic and Epigenetic Variation in Human Pluripotent Stem Cells

O. Kyriakides, J.A. Halliwell, and P.W. Andrews

Abstract Human pluripotent stem cells (hPSCs) can acquire non-random genomic variation during culture. Some of these changes are common in tumours and confer a selective growth advantage in culture. Additionally, there is evidence that reprogramming of human induced pluripotent stem cells (hiPSCs) introduces mutations. This poses a challenge to both the safety of clinical applications and the reliability of basic research using hPSCs carrying genomic variation. A number of methods are available for monitoring the genomic integrity of hPSCs, and a balance between practicality and sensitivity must be considered in choosing the appropriate methods for each use of hPSCs. Adjusting protocols by which hPSCs are derived and cultured is an evolving process that is important in minimising acquired genomic variation. Assessing genetic variation for its potential impact is becoming increasingly important as techniques to detect genome-wide variation improve.

Keywords Cytogenetics, Epigenetic, Genetic variants, Human, Karyotype, Pluripotent stem cells

Contents

O. Kyriakides, J.A. Halliwell, and P.W. Andrews (✉)
Centre for Stem Cell Biology, Department Biomedical Science, University of Sheffield, Western Bank, Sheffield S10 2TN, UK
e-mail: p.w.andrews@sheffield.ac.uk

Abbreviations

aCGH	array comparative genome hybridisation
CNV	Copy number variation
FISH	Fluorescent in situ hybridisation
hESC	Human embryonic stem cell
hiPSC	Human induced pluripotent stem cell
hPSC	Human pluripotent stem cell
NGS	Next-generation sequencing
qPCR	Quantitative polymerase chain reaction
SNP	Single nucleotide polymorphism
TGCT	Testicular germ cell tumour

1 Introduction

Human pluripotent stem cells (hPSCs) can be derived from embryos or induced from somatic cells [1–3]. These cells have the ability to produce cell types from any of the three germ layers and can self-renew. Excitement surrounding hPSCs is fuelled by potential uses in studying development, modelling disease and regenerative medicine.

Taking Parkinson's disease as an example, disease models have been developed by reprogramming patients' fibroblasts to human induced pluripotent stem cells (hiPSCs), facilitating a better understanding of the Parkinson's disease genotype [4]. Furthermore, by developing protocols for the differentiation of hPSCs to dopaminergic neurons, neuronal development cues have gradually become better understood [5]. This knowledge then allows the gradual translation into regenerative medicine treatments [6].

Similarly, using tissue from long QT patients, hiPSC-derived cardiac myocytes have been generated that show a characteristic reduction in the delayed rectifier potassium current [7]. Furthermore, the long QT hiPSC-derived cardiac myocyte model was used to screen for pharmacological agents providing an improvement to the phenotype [7].

These examples demonstrate the importance of hPSC research in a wide array of fields. To realise this potential, however, hPSCs must be maintained in culture, often in large numbers. hPSCs show apparent immortal self-renewal in culture, which distinguishes them from their in vivo embryonic counterparts, the fate of which quickly becomes restricted [8]. Since their first derivation, numerous studies have shown that hPSCs are subject to genomic change in culture. Furthermore, hiPSCs show additional signs of genetic instability associated with the reprogramming process.

This review first summarises current knowledge on acquired genomic change in hPSCs and then discusses emerging approaches for monitoring, minimising and assessing genomic change, which are important considerations in the field of hPSC research.

2   Genetic Change in Human Pluripotent Stem Cell Culture

Genetic changes can occur spontaneously in any cell but, through a combination of natural senescence and apoptosis, most never become established within the overall population. Indeed, post-mortem neural tissue shows low-level mosaicism, which may help to produce functional diversity [9]. Normal pluripotent stem cell populations are likewise chromosomally heterogeneous [10]. However, some hPSC cultures show non-random genetic changes that can come to dominate the population [11]. These commonly involve gains of parts of chromosomes 1, 12, 17 and 20 and losses of regions of chromosomes 10, 18 and 22 [12] (Fig. 1).

For example, in one study, over 50% of 30 human embryonic stem cell (hESC) lines maintained over 18 months developed karyotype abnormalities, with 17q and chromosome 12 trisomy being the most frequent changes [13]. Furthermore, the same karyotype abnormalities were reported independently in other lines [14, 15]. In all cases, the abnormalities were observed only after continued culture.

These changes are not exclusive to hESCs. A technique that infers karyotype abnormalities from gene expression data was used to demonstrate that genes on chromosome 12 were also consistently overexpressed in hiPSC lines [16]. Together,

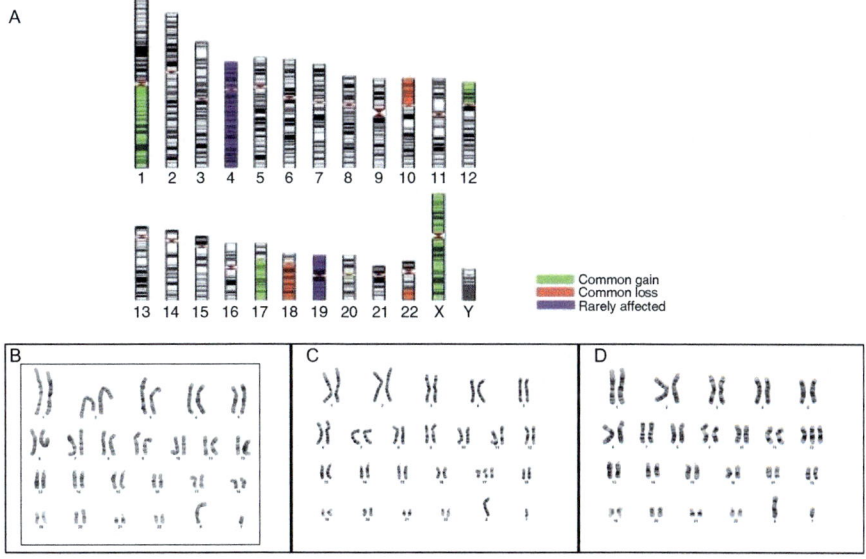

Fig. 1 Common abnormalities detected during the prolonged culture of hPSCs. (**a**) Ideogram depicting the commonly gained, lost and rarely affected chromosomes that are detected during prolonged culture of hPSCs [12]. (**b–d**) Examples of G banding karypotypes showing the gain of the long arm of chromosome 20 (**b**), gain of the whole of chromosome 17 (**c**), and gain of the whole of chromosome 12 (**d**). The gain of the long arm of chromosome 20 (**b**) has arisen as an isochromosome of 20q, with the consequent loss of the short arm of the chromosome. This particular cell is therefore trisomic for chromosome 20q but monosomic for chromosome 20p [12]

these data imply that the genetic aberrations observed are characteristic of pluripotent stem cell culture, rather than the source of the cells from either embryo or
fibroblast.

In a large scale screening of 125 hESC lines by single-nucleotide polymorphism
(SNP) array analysis, a sub-chromosomal copy number gain of part of the long arm
of chromosome 20 (20q11.21) was identified in 22 cell lines [12]. In all cases, the
duplications overlapped, sharing a minimal amplicon region of 0.55 Mb pairs. The
same copy number variant (CNV) has been identified independently in both hESCs
and hiPSCs [17].

Three genes within the 20q11.21 minimal amplicon are commonly expressed in
hESCs. One of which, *BCL2L1*, forms two alternative transcripts that encode both a
pro-apoptotic protein and an anti-apoptotic protein. In embryonic stem cells, the
anti-apoptotic protein BCL-XL is almost exclusively expressed [18].

This finding gives support to the hypothesis that the non-random genetic changes
observed within hPSCs are driven by selection, resulting in advantageous genetic
variation becoming widespread during long-term culture. Using the 20q11.21 CNV
as an example, we can assume this arises randomly and, because of the unlimited
proliferative potential of hPSCs, the extra dose of BCL-XL conferred by this CNV
could confer a selective advantage through its anti-apoptotic effects. Therefore,
continuous passaging of a cell culture carrying this CNV leads to its gradual
accumulation within the cell population.

Experimental evidence for this model was provided through comparison of
hESC lines carrying the 20q11.21 CNV with control hESC lines [18]. In this
study, population-doubling times of 35 and 138 h were reported, respectively.
Flow cytometry showed no difference in the distribution of cells throughout the
cell cycle within each population. Time-lapse confocal microscopy confirmed a
similar absolute cell division time. These results indicate that the reduced
population-doubling time observed in cells carrying 20q11.21 CNV was due to a
reduction in apoptosis rather than an increase in proliferation. The action of
BCL-XL specifically in this process was confirmed by overexpressing only
BCL-XL in a separate cell line, which mirrored the results of the cells carrying
the whole 20q11.21 CNV.

Strikingly, this process by which cells acquire a growth advantage during
prolonged culture closely resembles aspects of tumorigenesis (Fig. 2), which is
also thought to originate from mutations in a single cell that allow it to escape from
tight growth control, leading to selective clonal expansion [19]. It is therefore
possible that culture adaptation is an in vitro mimicry of this micro-evolutionary
process. This raises concerns for the clinical application of hPSCs because it is
plausible for such genetic change to confer malignant properties. For example, the
isochromosome of 12p is used as a clinical marker of testicular germ cell tumours
(TGCT) [20]. Furthermore, fluorescent in situ hybridisation (FISH) analysis of
human embryonal carcinoma cells (the malignant counterpart to hESCs) found
that 6/9 carried the 20q11.21 amplification [18], which suggests it can similarly
drive growth advantage in malignant cells.

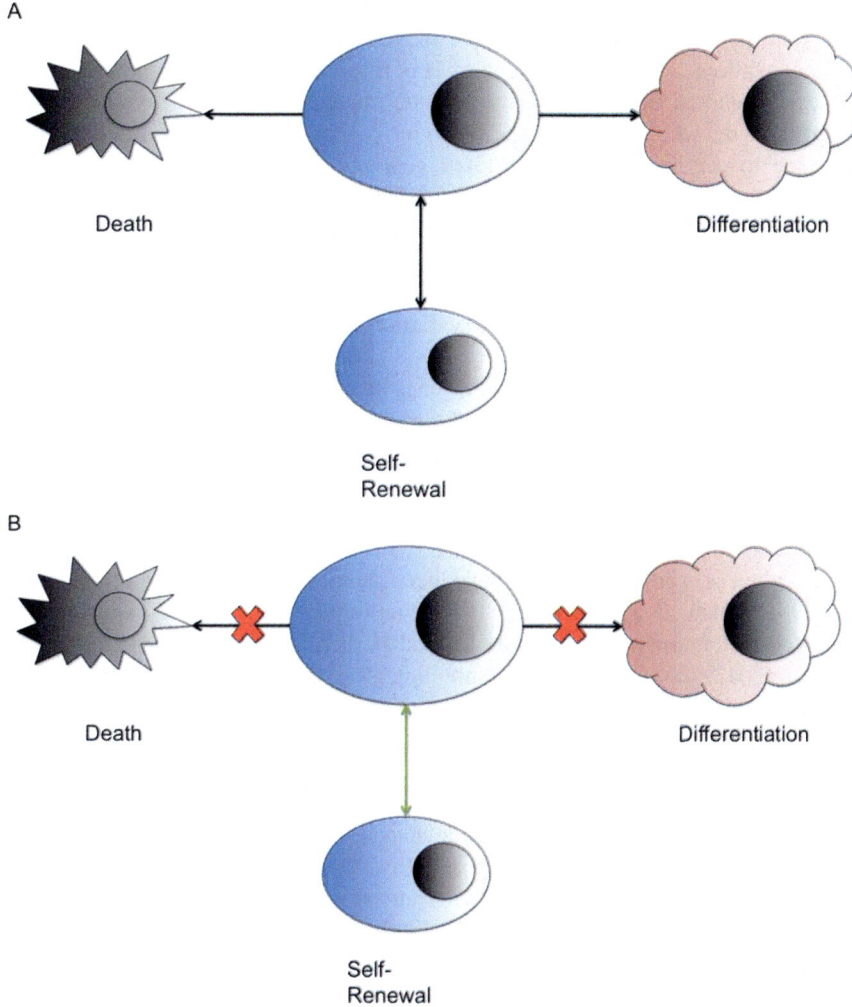

Fig. 2 Pluripotent stem cell fates. (**a**) Pluripotent stem cells have three main fate choices: undergo self-renewal producing two daughter stem cells, progress into differentiation (resulting in production of adult cell types), or undergo apoptosis. (**b**) Mutations that are advantageous to pluripotent stem cell fate include those that restrict their capacity to differentiate, inhibit cell death or enhance self-renewal

Genetic change has been detected on every chromosome during hPSC culture, although aberrations on chromosome 4 are exceptionally rare [12]. However, why particular aberrations, such as those on chromosomes 12, 17 and 20, are so common is still unknown. Recently, it was demonstrated that under replicative stress hESCs fail to activate key proteins (such as kinases CHK1 and ATR) involved in the S-phase checkpoint, despite normal levels of expression [21]. Furthermore, hESCs showed an upregulation in apoptotic markers and caspase 3 activation. This suggests that an intrinsic characteristic of hESCs is to eliminate cells with DNA damage, without an attempt at repair. This may be a desirable mechanism for protecting genome integrity because genetic change in ESCs in vivo would be passed on to the whole organism and could prove catastrophic. These findings have relevance to the discussion of acquired genomic variation. If hESCs normally protect genomic integrity through apoptosis rather than DNA repair, then an acquired variation such as the 20q11.21 CNV would provide a particular selective advantage. The resistance to apoptosis conferred by the extra dose of BCL-XL in cells carrying this CNV could help them thrive under these conditions. This could partly explain why chromosome 20q variations develop so commonly in hPSC cultures.

To date, similar evidence for driving genes on chromosomes 12 and 17 has been elusive. This is largely due to the scale of the changes. The 20q minimal amplicon is only 0.55 Mb, thus presenting a limited number of candidate genes to investigate. In contrast, the changes in chromosomes 12 and 17 usually involve a duplication of either the whole chromosome or an arm, so pinpointing the driving genes involved is more difficult.

Nevertheless, candidate genes have been suggested. For example, the gene *BIRC5*, located at 17q25.3, is known to have anti-apoptotic properties and is highly expressed in teratomas, the tumours formed by hESCs [22]. Likewise, *NANOG*, found at 12p13.31, contributes to maintaining pluripotency [23]. If overexpressed, *NANOG* may make cells more likely to continue self-renewal. However, detailed analysis shows that the closest minimal amplicon falls upstream of *NANOG* and includes its unexpressed pseudogene [17]. Furthermore, the same minimal amplicon was found to be just as prevalent in the reference samples [12] and, therefore, is unlikely to be the cause of a change in cell behaviour.

It is important not to dismiss the possibility that the phenotypic growth advantage conferred by these chromosomal aberrations is a result of a change in expression of multiple genes. This could explain why genetic change involving these chromosomes tends to involve whole or large duplications.

3   Epigenetic Change

The epigenetic status of a cell is highly important in gene expression and therefore in dictating its specific phenotype [24]. Particularly relevant are the processes of genome imprinting, whereby DNA methylation patterns produce monoallelic expression of particular genes in a parent-of-origin manner [25]. Previous studies observed epigenetic instability in cultured mouse ESCs [26] and hypothesised a link between assisted reproductive technology and epigenetic disorders [25]. This prompted investigation into whether removing hESCs from their in vivo environment and prolonged culture could perturb epigenetic imprinting.

In an early study of six imprinted genes in four hESC lines, the normally paternally imprinted gene *H19* gained biallelic expression during prolonged culture [27]. The *H19* gene stands out from the other five genes investigated because it acquires methylation during embryonic development. However, upon closer inspection, the re-expressed allele of *H19* still showed methylation typical of an imprinted gene, suggesting that re-expression occurs through an alternative mechanism [27].

In further studies of over 2,000 loci by restriction landmark genome scanning, all six hESC lines showed high levels of epigenetic instability, which was reliably fixed within the cell population [28]. Another study found that *IGF2* became biallelically expressed in an hESC line grown by one laboratory, whereas cultures of the same line grown by a different laboratory did not exhibit the same biallelic expression, which suggests that culture conditions can have an effect on the epigenetic status of cultured hESCs [29].

A more recent study of 205 hPSCs and 130 somatic samples provided interesting insights into tissue-specific versus pluripotent epigenetic character [30]. This study also detailed the correlation between either hypermethylation or hypomethylation with the loss of allele-specific expression of numerous genes in hPSCs. Additionally, the group reported that in female hPSCs, X chromosome inactivation was gradually lost with time in culture, corresponding to a decrease in *XIST* expression and an increase in mRNA expression of genes on this chromosome. This type of epigenetic instability is particularly relevant when considering the use of hPSCs in the modelling of X-linked diseases because it could confound results [31].

4 Further Considerations for Induced Pluripotent Stem Cells

The issues discussed regarding genetic and epigenetic change in culture are similar for both embryonic and induced pluripotent stem cells [32]. However, there are differences in hiPSCs that present further sources of genetic change in these cells.

hiPSCs differ from hESCs in that they are reprogrammed from somatic tissue. Originally, concerns were raised regarding the use of a retroviral vector for reprogramming [1] because integration of the transgene can produce insertion mutations, and insertional mutagenesis has previously been seen to cause serious adverse effects in a gene therapy attempt [33]. Attempts to address this issue include the development of reprogramming methods using an episome vector. This is able to replicate extrachromosomally, allowing reprogramming without integration. Furthermore, both vector and transgene can then be eliminated via drug selection [34].

Mutations possibly induced during reprogramming have been reported to occur in early passages of iPSCs [35], perhaps as a result of increased replicative stress caused by forced overexpression of reprogramming factors [36]. However, a comparison of hiPSC lines derived by retroviral or episomal reprogramming showed no significant difference in the frequency of karyotype abnormalities [32]. Detailed DNA sequence comparisons of parental somatic cells and hiPSCs derived from them indicated that many, if not all, of the mutations detected in the hiPSCs pre-existed in the parental somatic cells [37–39]. Because of the inefficiency of reprogramming, hiPSC lines usually have a clonal origin. Therefore, genetic change in just a single parental cell, not detectable in the bulk population because of limited sensitivity of the sequencing methods, could be carried through and mistakenly identified as a 'new' genetic variation when the hiPSC culture is compared with the parent culture as a whole [40]. Nevertheless, independent of 'mutations of origin', (i.e. those present in parental cells or induced during reprogramming), hiPSCs do tend to acquire the same common variants seen in hESCs during prolonged culture.

5   Monitoring Genetic Change

Monitoring hPSC cultures is important in the laboratory to ensure that genetic change does not affect experimental results. It is also vital in clinical applications to ensure that cells carrying potentially harmful genetic variations are not introduced into patients. A number of techniques are available to detect genetic change. Some methods screen the whole genome indiscriminately whereas others use probes targeted to known loci. The development of single-cell-based techniques makes it feasible to detect genetic change occurring in only a small minority of cells. All methods, however, have limitations and therefore judgement is required to ensure that hPSCs are monitored to an extent that is adequate for their use in either the laboratory or clinical setting.

The traditional, although still highly relevant, method for detecting genetic change in cell culture is by assessing the banding pattern of chromosomes in metaphase spreads. This was how some of the earliest genetic changes, such as those on chromosomes 12 and 17 were detected (Fig. 1) [11]. G-banding karyotype analysis has the advantage of allowing assessment of the whole genome for aberrations without any preconceived knowledge; however, it is highly labour intensive and analysis usually requires outsourcing to skilled cytogeneticists.

The process for G-banding involves preparing a certain number of metaphase spreads on a slide and scoring a random sample with the assumption that it is representative of the culture as a whole, although it is possible that differential growth patterns or detachment during harvesting of cells in mosaic cultures might distort this assumption. Recently, this assumption and the sensitivity of G-banding was tested systematically using mosaic cultures of hPSCs containing known genetic changes at increasing percentages within the population [41]. The results confirmed that acquired genetic change in hPSCs is detected by G-banding at the same frequency as statistically predicted using random sampling. However, sensitivity is limited by cost and practicalities. Typically, a cytogeneticist might score

Table 1 Ssensitivity of detecting karyotypically variant cells in mosaic cultures by G-banding karyology

Number of metaphases scored	Percentage of variant cells detected with 95% confidence (%)
20	28
30	18
50	13
60	10
100	6
500	<1

The table shows, based on statistical sampling theory, the minimum proportion of variant cells that would be detected in mosaic cultures for different numbers of metaphases scored [41]. By screening test cultures with different proportions of variant hESC, the actual sensitivity of G-banding karyology carried out using standard procedures closely matched the expected sensitivity predicted by statistical sampling theory

30 metaphases, but this will only reliably detect variants that are present in more than 18% of the cells in a mosaic culture (Table 1). A lower limit of around 6% mosaicism requires scoring 100 metaphase spreads. To detect variants present with less than 1% of the population requires screening over 500 metaphases, a number that is impracticable in routine cytogenetic practice.

G-banding karyotype analysis, even using newly developed automated techniques, is mostly restricted to detecting large genetic aberrations of over about 5 Mb [42]. Therefore, it is rare for small CNVs, such as the common 20q11.21 CNV, to be detected in this way. Typically, these require techniques such as single nucleotide polymorphism (SNP) array or array comparative genomic hybridisation (aCGH)-based analysis [12]. The potential of this CNV to be harmful is still unknown. However, as described, its anti-apoptotic property is known to confer a growth advantage and so any planned clinical application involving hPSCs should take account of the inability of karyotype analysis to detect this CNV.

Small CNVs such as that at 20q11.21 can also be detected using probe-based screening strategies, for example FISH. However, FISH suffers from many of the same issues as G-banding. It is labour intensive and has a limit of detection of around 5% due to false negatives. This is particularly an issue in the case of tandem duplications, when the signals from each copy may overlap and only one copy of the CNV is scored [41]. To overcome sampling issues, it is possible to combine FISH with flow cytometry in order to conduct a high-throughput screen. This interphase chromosome flow-FISH method has been tested on blood samples of myelodysplastic syndrome patients, who often present with chromosome 7 monosomy [43]. The study found that the technique reliably identified chromosome 7 monosomy without the need for laborious slide analysis. Automated flow cytometry also allows the screening of thousands of cells at once, making it less likely that a genetic aberration is undetected because of small sample size. Furthermore, the technique also provides a quantitative measure of the extent of aneuploidy in the sample.

Recently, a quantitative polymerase chain reaction (qPCR) method has been developed that allows detection of CNVs based on comparison of PCR products using primers selected for target and reference regions [41]. This technique was able to detect CNVs for chromosomes 12, 17 and 20 with a lower detection limit of 10%. This qPCR method provides a very useful technique for routinely checking laboratory cultures for known common genetic changes. However, both qPCR and FISH require pre-existing knowledge of genetic change in order to design primers or probes, respectively. This is probably not sufficient for clinical application because we do not yet know the full range of genetic change in hPSC culture or its ability to cause harm, and so a more unbiased screening method should also be employed.

Another powerful genome-wide screening method is SNP analysis, whereby CNVs are revealed by the increase or decrease in nearby SNP markers detected by microarray platforms. An alternative is aCGH, in which the comparison of samples to reference DNA is more integrated [44]. By hybridizing differentially probed reference and test samples to a microarray, the fluorescent ratios of each can be calculated. A ratio of 0 indicates normal or diploid condition, whereas ratios of -1 or $+1$ indicate a loss or gain, respectively, for that region. This technique has already proved powerful in the field of oncology [45]. Neither SNP array nor aCGH approaches require previous knowledge of the genetic change that might be present in a cell population. Furthermore, smaller CNVs of below 5 Mb in length can be detected by SNP arrays, with the resolution only limited by the distance between SNP markers. The usefulness of this technique has been demonstrated in the screening of 125 hESC lines, revealing that more than 20% carried 20q11.21 CNVs that were largely undetected by karyotype analysis [12]. However, although SNP-based techniques are more precise in terms of the size of CNV they can detect, they do not provide improved sensitivity in detecting CNVs present in only a minority of cells. In mixing tests, the ability of SNP microarray analysis to detect chromosome 8 trisomy became unreliable when it was present in less than 10% of the population [46]. Similar testing of aCGH revealed that the smallest CNVs were only detected in 10–15% of cultures [47]. Another limitation to SNP-based analysis is data interpretation. The sensitivity of the method for detecting small genetic changes means that numerous CNVs across all chromosomes are identified during screening [12]. However, the majority of these are stochastic in nature and do not produce a significant cellular phenotype. It is therefore a challenge to distinguish the relevant results from the background noise.

Next-generation sequencing (NGS) has revolutionised genome research and is increasingly used for the detection of structural variants. Many NGS approaches produce millions of short sequencing reads. By assuming that the distribution of these reads is random over the genome, it is possible to infer duplications or deletions from areas that do not follow this trend [48]. However, as with other techniques, NGS often fails to detect low-level variants of a mosaic population that are hidden by the normal signal. For example, in a study of tumour samples, a coverage as high as $10,000\times$ was required to confirm the presence of rare variants [49]. Drawing parallels from this highlights the difficulty of sensitively detecting low-level mosaicism in cultures of hPSCs. Also, sequencing of repetitive regions is still limited and sequencing or detection of the complete range of genetic variants may often require multiple strategies and sequencing approaches [48].

From these examples, it is apparent that there are numerous methods for monitoring genetic change in hPSCs. However, none alone fulfils all the requirements of a robust detection system. Karyotype analysis is still the best-validated and most widely used technology in clinical application [50] and detects large aberrations, such as those of chromosomes 12 and 17, but we know that significant small CNVs can be missed. Probes for well-characterised small CNVs, such as 20q11.21, allow FISH analysis to extend the range of known genetic change that can be detected, but this requires prior knowledge of the CNVs to be assayed. In laboratory applications, these techniques may be too labour intensive for routine assessment and so emerging techniques such as qPCR or interphase flow-FISH with a panel of primers or probes could allow screening for common genetic changes. It is important to recognise the limitations of detection methods, and judgement is required to achieve satisfactory monitoring of genetic change when using hPSCs in clinical or laboratory applications.

6   Minimising Genetic Change

Pertinent to the discussion of acquired genetic change in hPSCs are the measures that can be taken to reduce the rate at which genetic variants appear, recognising that their appearance depends upon two unrelated mechanisms, namely mutation and subsequent selection. Because much genetic change occurs through prolonged culture it is important to look closely at current methods of passaging and maintaining stem cells in culture. It is also important to discuss novel ways in which the mutation rate can be reduced and whether we can also reduce the selection pressure for potentially harmful genetic change.

Soon after karyotype abnormalities were first linked to prolonged hPSC culture, investigations into the possible effect of different passaging techniques were conducted. For example, one group showed that hESC lines could be maintained with a normal karyotype for prolonged periods using a manual passaging technique [15]. Furthermore, when these same lines were then switched to either enzymatic or non-enzymatic bulk passaging methods, characteristic genetic changes arose. A correlation between bulk passaging and karyotype abnormalities was documented in a large-scale screen [12]. Certainly, the correlation between bulk passaging methods and acquired genetic change may reflect the different stresses to which cells are exposed by different passaging techniques, but it may also reflect the greater number of cells that are transferred in bulk methods. For example, in a simulation study, the rate at which abnormal cells came to dominate the culture increased exponentially as the size of the overall cell population was increased [51]. This is probably a result of the greater number of cells undergoing individual mutational events, which increases the likelihood of a cell acquiring an advantageous change. This effect could partly explain the higher occurrence of acquired genetic change in hPSC cultures passaged by bulk methods, as the population size is greater.

The knowledge that population size affects the appearance of genetic variants in culture provides an opportunity to modify culturing methods. For example, another finding from the simulation studies by Olariu et al. [51] was that if the same number of cells was cultured in ten smaller subcultures, the rate at which abnormal cells appeared was lower than in one single large population. The maintenance of hPSCs in small subcultures could therefore be an effective way to minimise the effect of genetic change in culture. Furthermore, in the laboratory, if one subculture does acquire a significant level of genetic change then it can be easily discarded without abandoning the whole experiment. This could also be a useful consideration clinically because many potential regenerative medicine applications require a significant number of cells. Therefore, hPSCs could be expanded through many small subcultures before combining to produce the final treatment sample, although this may not be cost effective or practical for the needs of clinical scale-up.

Another consideration regarding hPSC maintenance is how much selection pressure is created by the culture method. It has been documented that a large amount of apoptosis occurs during the dissociation of hESC clumps during passaging [52] and it was estimated that roughly 90% of cells are lost between each passage [51]. This greatly increases the selection pressure for cells carrying a genetic change that confers a growth advantage. Increasing the efficiency with which cells are passaged would reduce this selection pressure and, therefore, reduce the occurrence of genetic change in culture. One study showed that a ROCK inhibitor could be used to reduce apoptosis during hESC dissociation, which significantly increased colony formation after cell transfer. In recent years, use of a ROCK inhibitor during hPSC passaging has become commonplace [52].

The predominant mechanism of mutation within hPSC culture is poorly understood, but studies have suggested novel ways to reduce the incidence of genetic change. For example, oxidative stress is widely implicated in DNA damage and hiPSCs have been documented to have levels of high reactive oxygen species (ROS) following reprogramming [53]. Furthermore, supplementing hiPSC cultures with antioxidants such as vitamin C reduced ROS levels and the cells had a reduced number of de novo CNVs [53]. The use of antioxidants would probably have a similar effect on the mutation rate in hESC culture.

Another possible approach is to use small molecule treatment to select against cells with different behaviour conferred by specific genetic variation. For example, one group demonstrated the increased sensitivity of hPSCs carrying trisomy of chromosome 12 to etoposide, cytarabine hydrochloride and gemcitabine hydrochloride [54], all DNA replication inhibitors already approved as anticancer therapies. Because many characterised hPSC genetic abnormalities confer a growth advantage, a similar strategy could be employed in culture to select against these cells.

7 Assessing the Effects of Genetic Change

Despite the possible avenues for reducing genetic change, it will be very difficult to culture hPSCs completely free of genetic alterations. Therefore, it is very important that we are able to assess genetic variants effectively to distinguish between the problematic and the harmless.

The well-documented abnormalities of chromosomes 12 and 17 can confer a growth advantage and, because of their large scale, cause aberrant expression of multiple genes in hPSCs [11]. Gene expression data from testicular germ cell tumours (TGCTs) show that copy number increases along chromosome 17q [55]. Isochromosome 12 is used as a clinical marker for TGCT [20]. Furthermore, investigators reported that a hESC line carrying chromosome 12 gains demonstrates neoplastic properties [56]. Together, these studies suggest that chromosome 12 and 17 abnormalities are unacceptable and, therefore, all clinical applications of hPSCs should require exclusion of these variants. Most clinical trials include G banding karyotype screens so that these large chromosomal abnormalities can be excluded with high confidence, providing a satisfactory number of metaphase spreads are analysed. However, the question as to when and how often clinically destined samples should be analysed is still unresolved.

Large genetic variation detected at the karyotype level is usually not acceptable for clinical use. However, a problem arises when considering smaller subchromosomal CNVs. The 20q11.21 CNV confers a growth advantage to cells in a similar manner to that associated with chromosome 12 and 17 abnormalities. Therefore, one would expect this to be a CNV that needs exclusion during clinical applications. Exclusion could be achieved using FISH analysis with a probe specific for the 20q11.21 region. Furthermore, a spectrum of probes could be developed to screen cells for known CNVs. However, genome-wide SNP analysis reveals a vast array of CNVs of a similar size to the 20q11.21 [12], but it is difficult to assess which of these may be harmful, either because they promote transformation and the development of cancer, or because they affect the function of the derivative cells to be used for therapy. In either case, the answer depends on the types of derivative cells produced. For example, the potential for converting non-dividing derivative cells such as cardiomyocytes to malignant derivatives is likely to be substantially less than for differentiated cells that still retain proliferative potential, such as hepatocytes.

Clues to the possible consequences for malignancy of genetically variant hPSC derivatives can be obtained from the various cancer genome databases that are now being developed, such as the International Cancer Genome Consortium (http://icgc. org/). However, direct assessment of malignant potential requires in vivo studies. In one study, investigators took a hESC line harbouring the 20q11.21 CNV with high proliferative capacity and growth factor independence [56]. They transplanted neural derivatives into mice where they formed tumours [57]. Similar studies testing other recurrent CNVs in vivo could help in the assessment of hPSC genetic variation.

Critical effects of genetic variants on cell function must be tailored to the specific cell types being produced, and could involve either in vivo or in vitro studies as appropriate. For example, a vital function of cardiomyocytes is their characteristic calcium handling, which has been used to compare hiPSC-derived cardiomyocytes to somatic cells [58]. Similar studies with hPSCs carrying a particular CNV could reveal whether the genetic variation disrupts the function of the specialized derivative. This would be extremely important for validating hPSCs as developmental and disease models.

8   Conclusion

Acquired genomic change is a concern for both its potential to confound the results of basic research and to jeopardise the safety of clinical applications. Despite this, trials using pluripotent stem cell products are in progress. The first such trial was launched by Geron in 2010 and aimed to use oligodendrocyte hESC derivatives to treat spinal cord injury. The study was discontinued in 2011 due to financial constraints, but a follow-up of the patients occurred at 3 years [59]. Cardiac progenitors from hESCs have also been used in a trial on heart failure [60]. A number of hESC-based trials for macular degeneration are also underway, including studies launched by Pfizer [61] and Ocata Therapeutics [62]. So far, no adverse effects relating to genomic change have been reported in any of these trials. However, it is important to remain vigilant.

Monitoring genetic change has different requirements for specific applications. In basic research, efficient and affordable methods are employed so that they can be applied routinely. Promising techniques, utilizing qPCR and flow cytometry, are therefore likely to be important developments. Monitoring genetic change for clinical applications is likewise changing. For example, in the earliest trial aimed at treating macular degeneration, although a normal karyotype was confirmed, further high-resolution techniques were not used [63]. However, in a more recent trial, FISH analysis using probes for loci on chromosomes 12, 17 and 20 was employed to screen for well-characterized changes associated with hPSC culture [60]. As our knowledge of genomic variation grows, additional probes can be added to this list to exclude other genetic changes. An argument can be made that the technology to screen the whole genome indiscriminately for single nucleotide variants and small CNVs is available in the form of NGS, aCGH and SNP analyses and should be used. However, defining what is a significant genetic change and what is part of normal variation is difficult.

As discussed, one method to assess the significance of genetic variation is through in vivo studies. This is a major step in bringing any stem-cell-based treatment to the clinic. The first macular degeneration trial was preceded by pre-clinical studies in 45 rats, which confirmed the safety of the hPSC-derived treatment in vivo 12 [64]. Because macular degeneration is a disease of the eye, the treatment area is relatively small. This meant that the same number of cells (5×10^4)

could be tested in the model as used in the human trial [63]. A problem that may arise when hPSC-based treatments are developed for larger organs is that the number of cells required will increase. Therefore, it may not be feasible to test the same number of cells in some model organisms because of the relative size of the organs. This is an important consideration because much acquired genetic change occurs during prolonged passage. Therefore, if pre-clinical trials are performed using a smaller number of cells, then it is possible that the extended culture time required to produce the required cell number in the human trial will introduce more genetic change.

As hPSCs continue to be used, it is likely that protocols will be adjusted to minimize genetic change. For example, a recently launched trial using hiPSC-derived retinal pigment epithelium to treat macular degeneration [65] was put on hold because of detection of a cancer-related mutation in the hiPSC sample [66]. This change was not detected in the original skin cells, so could either have been present at undetectable levels or caused by the reprogramming procedure [66]. The risk associated with this reprogramming technique could lead to increased movement towards non-integrative reprogramming techniques such as episomal vectors [34]. Splitting hPSC cultures into smaller subcultures, reducing selection pressure, and using antioxidants may also help to reduce the occurrence of acquired genetic change in culture.

Encouragement can be taken from the lack of adverse effects in human trials using hPSCs to date. However, it is imperative that this remains the case with future trials for both the safety of patients and to prevent stalling of hPSC applications. This aim will be aided by continual consideration of the monitoring, minimizing and assessing of genomic variation in the context of both basic research and clinical application.

Acknowledgement This project received funding from the European Union's Horizon 2020 research and innovation programme under grant agreement No. 668724.

European Research Council

Established by the European Commission

References

1. Takahashi K, Tanabe K, Ohnuki M, Narita M, Ichisaka T, Tomoda K, Yamanaka S (2007) Induction of pluripotent stem cells from adult human fibroblasts by defined factors. Cell 131:861–872
2. Thomson JA, Itskovitz-Eldor J, Shapiro SS, Waknitz MA, Swiergiel JJ, Marshall VS, Jones JM (1998) Embryonic stem cell lines derived from human blastocysts. Science 282:1145–1147
3. Yu J, Vodyanik MA, Smuga-Otto K, Antosiewicz-Bourget J, Frane JL, Tian S, Nie J, Jonsdottir GA, Ruotti V, Stewart R, Slukvin II, Thomson JA (2007) Induced pluripotent stem cell lines derived from human somatic cells. Science 318:1917–1920
4. Soldner F, Hockemeyer D, Beard C, Gao Q, Bell GW, Cook EG, Hargus G, Blak A, Cooper O, Mitalipova M, Isacson O, Jaenisch R (2009) Parkinson's disease patient-derived induced pluripotent stem cells free of viral reprogramming factors. Cell 136:964–977
5. Perrier AL, Tabar V, Barberi T, Rubio ME, Bruses J, Topf N, Harrison NL, Studer L (2004) Derivation of midbrain dopamine neurons from human embryonic stem cells. Proc Natl Acad Sci U S A 101:12543–12548
6. Kriks S, Shim JW, Piao J, Ganat YM, Wakeman DR, Xie Z, Carrillo-Reid L, Auyeung G, Antonacci C, Buch A, Yang L, Beal MF, Surmeier DJ, Kordower JH, Tabar V, Studer L (2011) Dopamine neurons derived from human ES cells efficiently engraft in animal models of Parkinson's disease. Nature 480:547–551
7. Itzhaki I, Maizels L, Huber I, Zwi-Dantsis L, Caspi O, Winterstern A, Feldman O, Gepstein A, Arbel G, Hammerman H, Boulos M, Gepstein L (2011) Modelling the long QT syndrome with induced pluripotent stem cells. Nature 471:225–229
8. Andrews PW (2002) From teratocarcinomas to embryonic stem cells. Philos Trans R Soc Lond Ser B Biol Sci 357:405–417
9. Rehen SK, Yung YC, Mccreight MP, Kaushal D, Yang AH, Almeida BS, Kingsbury MA, Cabral KM, Mcconnell MJ, Anliker B, Fontanoz M, Chun J (2005) Constitutional aneuploidy in the normal human brain. J Neurosci 25:2176–2180
10. Peterson SE, Westra JW, Rehen SK, Young H, Bushman DM, Paczkowski CM, Yung YC, Lynch CL, Tran HT, Nickey KS, Wang YC, Laurent LC, Loring JF, Carpenter MK, Chun J (2011) Normal human pluripotent stem cell lines exhibit pervasive mosaic aneuploidy. PLoS One 6:e23018
11. Draper JS, Smith K, Gokhale P, Moore HD, Maltby E, Johnson J, Meisner L, Zwaka TP, Thomson JA, Andrews PW (2004) Recurrent gain of chromosomes 17q and 12 in cultured human embryonic stem cells. Nat Biotechnol 22:53–54
12. Amps K, Andrews PW, Anyfantis G, Armstrong L, Avery S, Baharvand H, Baker J, Baker D, Munoz MB, Beil S, Benvenisty N, Ben-Yosef D, Biancotti JC, Bosman A, Brena RM, Brison D, Caisander G, Camarasa MV, Chen J, Chiao E, Choi YM, Choo AB, Collins D, Colman A, Crook JM, Daley GQ, Dalton A, de Sousa PA, Denning C, Downie J, Dvorak P, Montgomery KD, Feki A, Ford A, Fox V, Fraga AM, Frumkin T, Ge L, Gokhale PJ, Golan-Lev T, Gourabi H, Gropp M, Lu G, Hampl A, Harron K, Healy L, Herath W, Holm F, Hovatta O, Hyllner J, Inamdar MS, Irwanto AK, Ishii T, Jaconi M, Jin Y, Kimber S, Kiselev S, Knowles BB, Kopper O, Kukharenko V, Kuliev A, Lagarkova MA, Laird PW, Lako M, Laslett AL, Lavon N, Lee DR, Lee JE, Li C, Lim LS, Ludwig TE, Ma Y, Maltby E, Mateizel I, Mayshar Y, Mileikovsky M, Minger SL, Miyazaki T, Moon SY, Moore H, Mummery C, Nagy A, Nakatsuji N, Narwani K, Oh SK, Olson C, Otonkoski T, Pan F, Park IH, Pells S, Pera MF, Pereira LV, Qi O, Raj GS, Reubinoff B, Robins A, Robson P, Rossant J, Salekdeh GH, Schulz TC et al (2011) Screening ethnically diverse human embryonic stem cells identifies a chromosome 20 minimal amplicon conferring growth advantage. Nat Biotechnol 29:1132–1144
13. Baker DE, Harrison NJ, Maltby E, Smith K, Moore HD, Shaw PJ, Heath PR, Holden H, Andrews PW (2007) Adaptation to culture of human embryonic stem cells and oncogenesis in vivo. Nat Biotechnol 25:207–215

14. Cowan CA, Klimanskaya I, Mcmahon J, Atienza J, Witmyer J, Zucker JP, Wang S, Morton CC, Mcmahon AP, Powers D, Melton DA (2004) Derivation of embryonic stem-cell lines from human blastocysts. N Engl J Med 350:1353–1356
15. Mitalipova MM, Rao RR, Hoyer DM, Johnson JA, Meisner LF, Jones KL, Dalton S, Stice SL (2005) Preserving the genetic integrity of human embryonic stem cells. Nat Biotechnol 23:19–20
16. Mayshar Y, Ben-David U, Lavon N, Biancotti JC, Yakir B, Clark AT, Plath K, Lowry WE, Benvenisty N (2010) Identification and classification of chromosomal aberrations in human induced pluripotent stem cells. Cell Stem Cell 7:521–531
17. Laurent LC, Ulitsky I, Slavin I, Tran H, Schork A, Morey R, Lynch C, Harness JV, Lee S, Barrero MJ, Ku S, Martynova M, Semechkin R, Galat V, Gottesfeld J, Izpisua Belmonte JC, Murry C, Keirstead HS, Park HS, Schmidt U, Laslett AL, Muller FJ, Nievergelt CM, Shamir R, Loring JF (2011) Dynamic changes in the copy number of pluripotency and cell proliferation genes in human ESCs and iPSCs during reprogramming and time in culture. Cell Stem Cell 8:106–118
18. Avery S, Hirst AJ, Baker D, Lim CY, Alagaratnam S, Skotheim RI, Lothe RA, Pera MF, Colman A, Robson P, Andrews PW, Knowles BB (2013) BCL-XL mediates the strong selective advantage of a 20q11.21 amplification commonly found in human embryonic stem cell cultures. Stem Cell Rep 1:379–386
19. Fialkow PJ, Jacobson RJ, Papayannopoulou T (1977) Chronic myelocytic leukemia: clonal origin in a stem cell common to the granulocyte, erythrocyte, platelet and monocyte/macrophage. Am J Med 63:125–130
20. Bosl GJ, Dmitrovsky E, Reuter VE, Samaniego F, Rodriguez E, Geller NL, Chaganti RS (1989) Isochromosome of the short arm of chromosome 12: clinically useful markers for male germ cell tumors. J Natl Cancer Inst 81:1874–1878
21. Desmarais JA, Hoffmann MJ, Bingham G, Gagou ME, Meuth M, Andrews PW (2012) Human embryonic stem cells fail to activate CHK1 and commit to apoptosis in response to DNA replication stress. Stem Cells 30:1385–1393
22. Blum B, Bar-Nur O, Golan-Lev T, Benvenisty N (2009) The anti-apoptotic gene survivin contributes to teratoma formation by human embryonic stem cells. Nat Biotechnol 27:281–287
23. Mitsui K, Tokuzawa Y, Itoh H, Segawa K, Murakami M, Takahashi K, Maruyama M, Maeda M, Yamanaka S (2003) The homeoprotein Nanog is required for maintenance of pluripotency in mouse epiblast and ES cells. Cell 113:631–642
24. Ferguson-Smith AC (2011) Genomic imprinting: the emergence of an epigenetic paradigm. Nat Rev Genet 12:565–575
25. Butler MG (2009) Genomic imprinting disorders in humans: a mini-review. J Assist Reprod Genet 26:477–486
26. Humpherys D, Eggan K, Akutsu H, Hochedlinger K, Rideout WM, Biniszkiewicz D, Yanagimachi R, Jaenisch R (2001) Epigenetic instability in ES cells and cloned mice. Science 293:95–97
27. Rugg-Gunn PJ, Ferguson-Smith AC, Pedersen RA (2005) Epigenetic status of human embryonic stem cells. Nat Genet 37:585–587
28. Allegrucci C, Wu YZ, Thurston A, Denning CN, Priddle H, Mummery CL, Ward-van Oostwaard D, Andrews PW, Stojkovic M, Smith N, Parkin T, Jones ME, Warren G, Yu L, Brena RM, Plass C, Young LE (2007) Restriction landmark genome scanning identifies culture-induced DNA methylation instability in the human embryonic stem cell epigenome. Hum Mol Genet 16:1253–1268
29. Rugg-Gunn PJ, Ferguson-Smith AC, Pedersen RA (2007) Status of genomic imprinting in human embryonic stem cells as revealed by a large cohort of independently derived and maintained lines. Hum Mol Genet 16(Spec 2):R243–R251
30. Nazor KL, Altun G, Lynch C, Tran H, Harness JV, Slavin I, Garitaonandia I, Müller FJ, Wang YC, Boscolo FS, Fakunle E, Dumevska B, Lee S, Park HS, Olee T, D'lima DD, Semechkin R, Parast MM, Galat V, Laslett AL, Schmidt U, Keirstead HS, Loring JF, Laurent LC (2012) Recurrent variations in DNA methylation in human pluripotent stem cells and their differentiated derivatives. Cell Stem Cell 10:620–634

31. Mekhoubad S, Bock C, De Boer AS, Kiskinis E, Meissner A, Eggan K (2012) Erosion of dosage compensation impacts human iPSC disease modeling. Cell Stem Cell 10:595–609
32. Taapken SM, Nisler BS, Newton MA, Sampsell-Barron TL, Leonhard KA, Mcintire EM, Montgomery KD (2011) Karotypic abnormalities in human induced pluripotent stem cells and embryonic stem cells. Nat Biotechnol 29:313–314
33. Hacein-Bey-Abina S, Von Kalle C, Schmidt M, Mccormack MP, Wulffraat N, Leboulch P, Lim A, Osborne CS, Pawliuk R, Morillon E, Sorensen R, Forster A, Fraser P, Cohen JI, de Saint Basile G, Alexander I, Wintergerst U, Frebourg T, Aurias A, Stoppa-Lyonnet D, Romana S, Radford-Weiss I, Gross F, Valensi F, Delabesse E, Macintyre E, Sigaux F, Soulier J, Leiva LE, Wissler M, Prinz C, Rabbitts TH, Le Deist F, Fischer A, Cavazzana-Calvo M (2003) LMO2-associated clonal T cell proliferation in two patients after gene therapy for SCID-X1. Science 302:415–419
34. Yu J, Hu K, Smuga-Otto K, Tian S, Stewart R, Slukvin II, Thomson JA (2009) Human induced pluripotent stem cells free of vector and transgene sequences. Science 324:797–801
35. Hussein SM, Batada NN, Vuoristo S, Ching RW, Autio R, Narva E, Ng S, Sourour M, Hamalainen R, Olsson C, Lundin K, Mikkola M, Trokovic R, Peitz M, Brustle O, Bazett-Jones DP, Alitalo K, Lahesmaa R, Nagy A, Otonkoski T (2011) Copy number variation and selection during reprogramming to pluripotency. Nature 471:58–62
36. Ruiz S, Lopez-Contreras AJ, Gabut M, Marion RM, Gutierrez-Martinez P, Bua S, Ramirez O, Olalde I, Rodrigo-Perez S, Li H, Marques-Bonet T, Serrano M, Blasco MA, Batada NN, Fernandez-Capetillo O (2015) Limiting replication stress during somatic cell reprogramming reduces genomic instability in induced pluripotent stem cells. Nat Commun 6:8036
37. Gore A, Li Z, Fung HL, Young JE, Agarwal S, Antosiewicz-Bourget J, Canto I, Giorgetti A, Israel MA, Kiskinis E, Lee JH, Loh YH, Manos PD, Montserrat N, Panopoulos AD, Ruiz S, Wilbert ML, Yu J, Kirkness EF, Izpisua Belmonte JC, Rossi DJ, Thomson JA, Eggan K, Daley GQ, Goldstein LS, Zhang K (2011) Somatic coding mutations in human induced pluripotent stem cells. Nature 471:63–67
38. Ji J, Ng SH, Sharma V, Neculai D, Hussein S, Sam M, Trinh Q, Church GM, Mcpherson JD, Nagy A, Batada NN (2012) Elevated coding mutation rate during the reprogramming of human somatic cells into induced pluripotent stem cells. Stem Cells 30:435–440
39. Rouhani FJ, Nik-Zainal S, Wuster A, Li Y, Conte N, Koike-Yusa H, Kumasaka N, Vallier L, Yusa K, Bradley A (2016) Mutational history of a human cell lineage from somatic to induced pluripotent stem cells. PLoS Genet 12:e1005932
40. Young MA, Larson DE, Sun CW, George DR, Ding L, Miller CA, Lin L, Pawlik KM, Chen K, Fan X, Schmidt H, Kalicki-Veizer J, Cook LL, Swift GW, Demeter RT, Wendl MC, Sands MS, Mardis ER, Wilson RK, Townes TM, Ley TJ (2012) Background mutations in parental cells account for most of the genetic heterogeneity of induced pluripotent stem cells. Cell Stem Cell 10:570–582
41. Baker D, Hirst AJ, Gokhale PJ, Juarez MA, Williams S, Wheeler M, Bean K, Allison TF, Moore HD, Andrews PW, Barbaric I (2016) Detecting genetic mosaicism in cultures of human pluripotent stem cells. Stem Cell Rep 7:998–1012
42. Steinemann D, Göhring G, Schlegelberger B (2013) Genetic instability of modified stem cells - a first step towards malignant transformation? Am J Stem Cells 2:39–51
43. Keyvanfar K, Weed J, Swamy P, Kajigaya S, Calado RT, Young NS (2012) Interphase chromosome flow-FISH. Blood 120:e54–e59
44. Rassekh SR, Chan S, Harvard C, Dix D, Qiao Y, Rajcan-Separovic E (2008) Screening for submicroscopic chromosomal rearrangements in Wilms tumor using whole-genome microarrays. Cancer Genet Cytogenet 182:84–94
45. Pinkel D, Albertson DG (2005) Array comparative genomic hybridization and its applications in cancer. Nat Genet 37(Suppl):S11–S17
46. Cross J, Peters G, Wu Z, Brohede J, Hannan GN (2007) Resolution of trisomic mosaicism in prenatal diagnosis: estimated performance of a 50K SNP microarray. Prenat Diagn 27:1197–1204

47. Valli R, Marletta C, Pressato B, Montalbano G, Lo Curto F, Pasquali F, Maserati E (2011) Comparative genomic hybridization on microarray (a-CGH) in constitutional and acquired mosaicism may detect as low as 8% abnormal cells. Mol Cytogenet 4:13
48. Tattini L, D'aurizio R, Magi A (2015) Detection of genomic structural variants from next-generation sequencing data. Front Bioeng Biotechnol 3:92
49. Griffith M, Miller CA, Griffith OL, Krysiak K, Skidmore ZL, Ramu A, Walker JR, Dang HX, Trani L, Larson DE, Demeter RT, Wendl MC, McMichael JF, Austin RE, Magrini V, McGrath SD, Ly A, Kulkarni S, Cordes MG, Fronick CC, Fulton RS, Maher CA, Ding L, Klco JM, Mardis ER, Ley TJ, Wilson RK (2015) Optimizing cancer genome sequencing and analysis. Cell Syst 1:210–223
50. Whiting P, Kerby J, Coffey P, Da Cruz L, Mckernan R (2015) Progressing a human embryonic stem-cell-based regenerative medicine therapy towards the clinic. Philos Trans R Soc London Ser B Biol Sci 370:20140375
51. Olariu V, Harrison NJ, Coca D, Gokhale PJ, Baker D, Billings S, Kadirkamanathan V, Andrews PW (2010) Modeling the evolution of culture-adapted human embryonic stem cells. Stem Cell Res 4:50–56
52. Watanabe K, Ueno M, Kamiya D, Nishiyama A, Matsumura M, Wataya T, Takahashi JB, Nishikawa S, Muguruma K, Sasai Y (2007) A ROCK inhibitor permits survival of dissociated human embryonic stem cells. Nat Biotechnol 25:681–686
53. Ji J, Sharma V, Qi S, Guarch ME, Zhao P, Luo Z, Fan W, Wang Y, Mbabaali F, Neculai D, Esteban MA, Mcpherson JD, Batada NN (2014) Antioxidant supplementation reduces genomic aberrations in human induced pluripotent stem cells. Stem Cell Rep 2:44–51
54. Ben-David U, Arad G, Weissbein U, Mandefro B, Maimon A, Golan-Lev T, Narwani K, Clark AT, Andrews PW, Benvenisty N, Carlos Biancotti J (2014) Aneuploidy induces profound changes in gene expression, proliferation and tumorigenicity of human pluripotent stem cells. Nat Commun 5:4825
55. Skotheim RI, Monni O, Mousses S, Fosså SD, Kallioniemi OP, Lothe RA, Kallioniemi A (2002) New insights into testicular germ cell tumorigenesis from gene expression profiling. Cancer Res 62:2359–2364
56. Werbowetski-Ogilvie TE, Bossé M, Stewart M, Schnerch A, Ramos-Mejia V, Rouleau A, Wynder T, Smith MJ, Dingwall S, Carter T, Williams C, Harris C, Dolling J, Wynder C, Boreham D, Bhatia M (2009) Characterization of human embryonic stem cells with features of neoplastic progression. Nat Biotechnol 27:91–97
57. Werbowetski-Ogilvie TE, Morrison LC, Fiebig-Comyn A, Bhatia M (2012) In vivo generation of neural tumors from neoplastic pluripotent stem cells models early human pediatric brain tumor formation. Stem Cells 30:392–404
58. Hwang HS, Kryshtal DO, Feaster TK, Sánchez-Freire V, Zhang J, Kamp TJ, Hong CC, Wu JC, Knollmann BC (2015) Comparable calcium handling of human iPSC-derived cardiomyocytes generated by multiple laboratories. J Mol Cell Cardiol 85:79–88
59. Asterias B (2014) Asteria Biotherapeutics, Inc. Announces new results from first-in-man clinical trial of a cell therapy derived from embryonic stem cells [Online]. http://asteriasbiotherapeutics.com/asterias-biotherapeutics-inc-announces-new-results-from-first-in-man-clinical-trial-of-a-cell-therapy-derived-from-embryonic-stem-cells/:   Asteria Biotherapeutics. Accessed May 22, 2014 2015
60. Menasché P, Vanneaux V, Hagège A, Bel A, Cholley B, Cacciapuoti I, Parouchev A, Benhamouda N, Tachdjian G, Tosca L, Trouvin JH, Fabreguettes JR, Bellamy V, Guillemain R, Suberbielle Boissel C, Tartour E, Desnos M, Larghero J (2015) Human embryonic stem cell-derived cardiac progenitors for severe heart failure treatment: first clinical case report. Eur Heart J 36:2011–2017
61. Pfizer & London, U. C. 2012. https://clinicaltrials.gov/ct2/show/NCT01691261: Clinical Trials. Accessed 22 Sept 2012
62. Schwartz SD, Regillo CD, Lam BL, Eliott D, Rosenfeld PJ, Gregori NZ, Hubschman JP, Davis JL, Heilwell G, Spirn M, Maguire J, Gay R, Bateman J, Ostrick RM, Morris D, Vincent M,

Anglade E, Del Priore LV, Lanza R (2015) Human embryonic stem cell-derived retinal pigment epithelium in patients with age-related macular degeneration and Stargardt's macular dystrophy: follow-up of two open-label phase 1/2 studies. Lancet 385:509–516

63. Schwartz SD, Hubschman JP, Heilwell G, Franco-Cardenas V, Pan CK, Ostrick RM, Mickunas E, Gay R, Klimanskaya I, Lanza R (2012) Embryonic stem cell trials for macular degeneration: a preliminary report. Lancet 379:713–720

64. Li Y, Tsai YT, Hsu CW, Erol D, Yang J, Wu WH, Davis RJ, Egli D, Tsang SH (2012) Long-term safety and efficacy of human-induced pluripotent stem cell (iPS) grafts in a preclinical model of retinitis pigmentosa. Mol Med 18:1312–1319

65. David C (2014) Japanese woman is first recipient of next-generation stem cells. Nature News: Nature Publishing Group, London

66. Andy C (2015) Mutation alert halts stem cell trial to cure blindness. New Scientist: New Scientist

Adv Biochem Eng Biotechnol (2018) 163: 207–220
DOI: 10.1007/10_2017_23
© Springer International Publishing AG 2017
Published online: 26 October 2017

Requirements for Using iPSC-Based Cell Models for Assay Development in Drug Discovery

Klaus Christensen, Filip Roudnicky, Christoph Patsch, and Mark Burcin

Abstract A prevalent challenge in drug discovery is the translation of findings from preclinical research into clinical success. Currently, more physiological in vitro systems are being developed to overcome some of these challenges. In particular, induced pluripotent stem cells (iPSCs) have provided the opportunity to generate human cell types that can be utilized for developing more disease-relevant cellular assay models. As the use of these complex models is lengthy and fairly complicated, we lay out our experiences of the cultivation, differentiation, and quality control requirements to successfully utilize pluripotent stem cells in drug discovery.

Keywords Assay development, Disease models, Drug discovery, Genomic integrity stability, Human pluripotent stem cells, Induced pluripotent stem cells, Quality control requirements

Contents

K. Christensen, F. Roudnicky, C. Patsch, and M. Burcin (✉)
Roche pRED (Pharmaceutical Research and Early Development), Roche Innovation Center
Basel, F.Hoffmann-La Roche Ltd., Basel, Switzerland
e-mail: mark.burcin@roche.com

1 Introduction

Over the last decade, the pharmaceutical industry has been changing the way drug development is conducted. Traditionally, drugs were developed in a linear approach starting with identification of biological targets that play a key role in the onset or progression of a given disease. This step was followed by a target identification phase, regularly using high-throughput screening to identify therapeutic agents that can modulate the target's activity. The efficacy and safety of these therapeutic agents were further optimized using different assays and models in vivo. Throughout the years, this approach has identified few candidates that have been developed into approved drugs [1]. As a consequence, the pharmaceutical industry is trying to move away from this linear approach in drug discovery to a more parallel system, whereby multiple assay formats are used from the beginning to identify several starting points in the process of drug development [2]. In addition, and because of the complexity of diseases, there is a need for deeper understanding of biological pathways and the molecular basis of human diseases. Furthermore, many efforts to develop disease models have provided a potentially higher predictability of identifying efficacious drugs. The availability of better disease models using suitable cell assays gives the opportunity to screen drug candidates that cause a desirable disease-reverting change in the phenotype.

The development of disease-related cellular models that mimic in vivo physiological processes is challenging and requires novel technologies. At present, many functional assays use cell culture models derived from primary tissue or artificially immortalized cell types. Such models have been widely used because of their expandability and ease of handling and cultivation. However, the cells used in such models are often not of human origin and may carry karyotype abnormalities. These anomalies often result in phenotypic changes and cause the disease relevance of such models to be questioned [3].

Primary cell cultures offer an alternative to recombinant cell lines; however, their use in large drug screening campaigns is restricted because of difficulties in generating large quantities of primary cells. In addition, donor variability makes it difficult to generate robust and reproducible results in compliance with drug development regulations.

The use of pluripotent stem cells (PSCs) by the pharmaceutical industry for the development of cellular assays for drug discovery is moving to center stage. The isolation of human PSCs from the inner cell mass of a blastocyst (known as human embryonic stem cells; hESCs) provides the opportunity to expand this cell source indefinitely. In addition, their pluripotent nature suggests that they could be used to derive many, if not all, differentiated cell types of the human body [4]. Considering this, various efforts have been made to implement PSC-based technologies for drug screening. This includes efforts to identify conditions that induce PSC differentiation to specific cell types [5–7]. The ability to cultivate, expand, and bank pluripotent and progenitor cells using defined culture media conditions is an important feature for the use of cells in drug discovery. The PSC field received increased attention in 2006, when Shinya Yamanaka demonstrated that PSCs could be directly reprogrammed from adult cells [8] and named these cells "induced pluripotent stem cells" (iPSCs). The advantage of iPSCs is that they do not carry the ethical burden of hESCs, which are derived from surplus human embryos. Furthermore, they allow an "individualized medicine" approach because they can be derived from individual patients. For inherited disorders, the ability to generate iPSCs [9] further extends the opportunity to develop patient-specific cell models carrying genetic mutations responsible for a defined disease. By using such "diseased" cell types, the effect of a specific mutation can be better understood in a pathway-centric view.

Genome editing is another recent breakthrough technology that further enhances our ability to develop customized cell assays for drug discovery. Genome editing is accomplished by using reverse genetics starting from particular genotypes, and then analyzing the resultant phenotypes. Furthermore, genome editing techniques in PSCs allow, for the first time, the use of this approach in models relevant to human disease. Genome editing in PSCs is used to make targeted modifications to the genome, either by generating loss-of-function alleles or by recreating monogenic disease mutations already known [10]. Efficient genome editing in human PSCs can be achieved by using either zinc finger nucleases (ZFNs), engineered transcription activator-like effector nucleases (TALENs), or the recently developed clustered regularly interspaced short palindromic repeat (CRISPR) technology. The CRISPR/Cas (CRISPR-associate endonuclease) system has been used for efficient genome editing in PSCs and to generate homozygous mutant clones readily [11, 12]. The combination of genome editing and iPSC technologies has granted the scientific community the opportunity to derive disease-relevant cellular models. Furthermore, by using genome editing technologies it is possible to revert the disease-causing mutations in iPSCs to obtain isogenic healthy control cells. For phenotypic screening efforts, isogenic iPSC control lines provide important information on whether the identified phenotype was indeed caused by the specific patient-specific mutation.

Traditionally, high-throughput screening campaigns have been the focus of many drug discovery programs. However, incorporation of PSCs into high-throughput screening presents diverse challenges, such as automated cell handling, miniaturization, and well-to-well variability. Therefore, the use of PSC-derived disease-relevant assays is still limited within high-throughput screening campaigns. However, the highest impact to the pharmaceutical industry is provided by PSC models, with secondary and more focused screening campaigns. This is especially evident during hit validation, qualification, and profiling. Here, disease-relevant assays aid in the identification of high quality drug candidates with the potential to translate into clinical efficacy. These human PSC models are especially important in cases where there is a need to identify human-specific modalities such as antibodies or RNA therapeutics. In such cases, it becomes difficult and often impossible to work with non-human drug discovery models.

Upon initiation of a drug discovery project, the biology of a specific target and its association with disease development is often not fully understood. In these cases, disease-relevant cellular assays are used to further validate a target and provide better insights into the mechanistic pathways of a specific disease. In addition, during the last decade, more and more phenotypic screens have led to drug candidates with the ability to revert a disease phenotype without knowing the target. Here, the use of PSC assays in combination with genome editing is a powerful tool for deciphering the target of a drug candidate during target identification.

The pharmaceutical industry undertakes strong efforts to avoid unwanted toxic side effects from drug candidates. The primary objective of such toxicology studies in the drug development process is to evaluate the safety of potential drug candidates. PSC-derived models are becoming widely used to identify such side effects more reliably and as early as possible during the process of drug development.

The following sections of this chapter summarize our experiences of the cultivation, differentiation, and quality control requirements for successfully utilizing PSCs in drug discovery.

2 Human PSC Cultivation Methods and Quality Control Requirements

Conventionally, hESCs have been cultured on feeder cells since the first embryonic stem cell cultures were established [4] (Fig. 1a). The most commonly used feeder cells are mouse embryonic fibroblasts. Feeder cells provide an optimal microenvironment for undifferentiated growth because of the presence of extracellular matrix components and growth factors, which are not yet fully understood and therefore bring additional experimental variability. The maintenance of these cultures requires a significant commitment of time and resources. Establishing the culture might take weeks and, once established, the cultures require care and replacement of culture media on a daily basis. During the expansion of PSCs, spontaneously differentiating areas can be observed by microscopy and then have to be removed. Instead of manual manipulation, such differentiated cell clusters can be automatically removed using the laser system PALM from Zeiss [13] or with specific detachment reagents. Development of feeder layer-free cultivation methods [14] has improved upscaling of the cultures and enhanced the reproducibility of differentiation protocols. To identify the best possible culture conditions for PSCs, our laboratory has tested several feeder-free media in combination with different matrices for support of cell attachment and growth. Because different PSC lines have different culture requirements, our efforts have not identified a "one-fits-all" culture method that provides optimal conditions for all cell lines tested. For this reason, different PSC lines require individual attention and adaptation of their cultivation method, making this a labor-intensive effort.

Fig. 1 Hallmarks of pluripotent stem cells. (**a**) Image of typical morphology of PSC on feeder- and feeder-layer free conditions. (**b**) Immunofluorescence stainings of pluripotency associated markers. (**c**) Flow cytometric analysis using TRA1-60 and SSEA1 specific antibodies

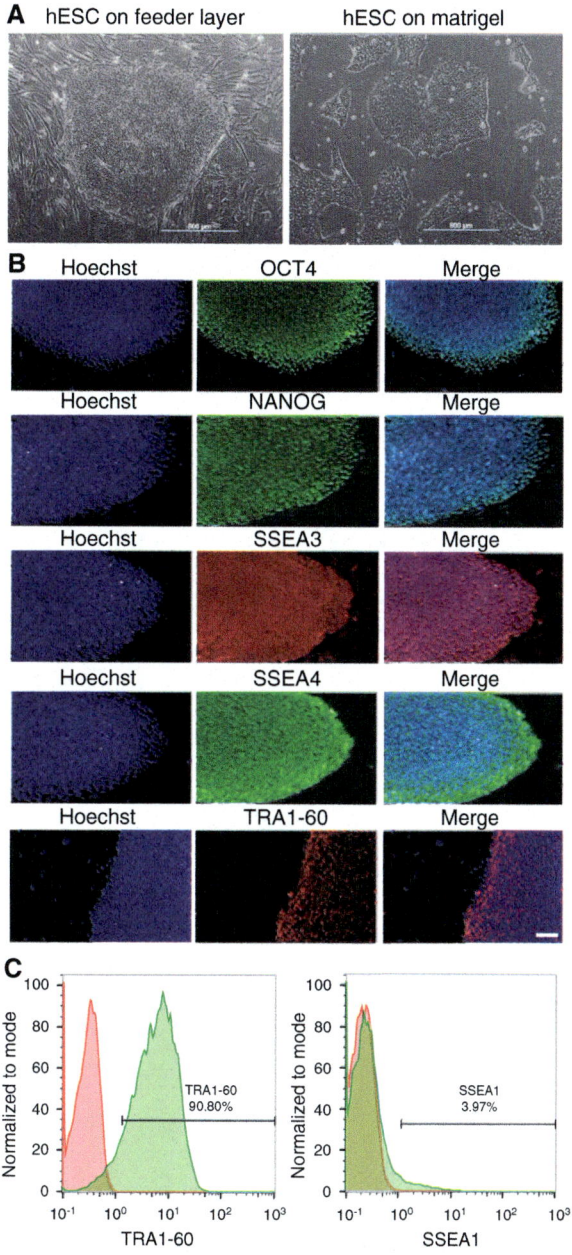

To ensure the highest level of consistency and reproducibility when using PSC models in drug discovery assays, we have implemented a number of quality control parameters, as described in Sects. 2.1–2.4.

2.1 Microbial Contamination Testing

Bacteria and fungi are common environmental contaminants that can infect cell cultures and make them unusable. To detect bacterial contamination as early as possible we do not use any antibiotics or antimycotics in human PSC cultures. Under these conditions, bacteria or fungi are not suppressed by antibiotics or antimycotics. The most common organisms known to cause unrecognized contamination are *Mycoplasma* and *Acholeplasma* spp. [15, 16]. For this reason, our cultures are tested on a regular basis for *Mycoplasma/Acholeplasma* using specific detection methods such as the polymerase chain reaction (PCR; Fig. 2a).

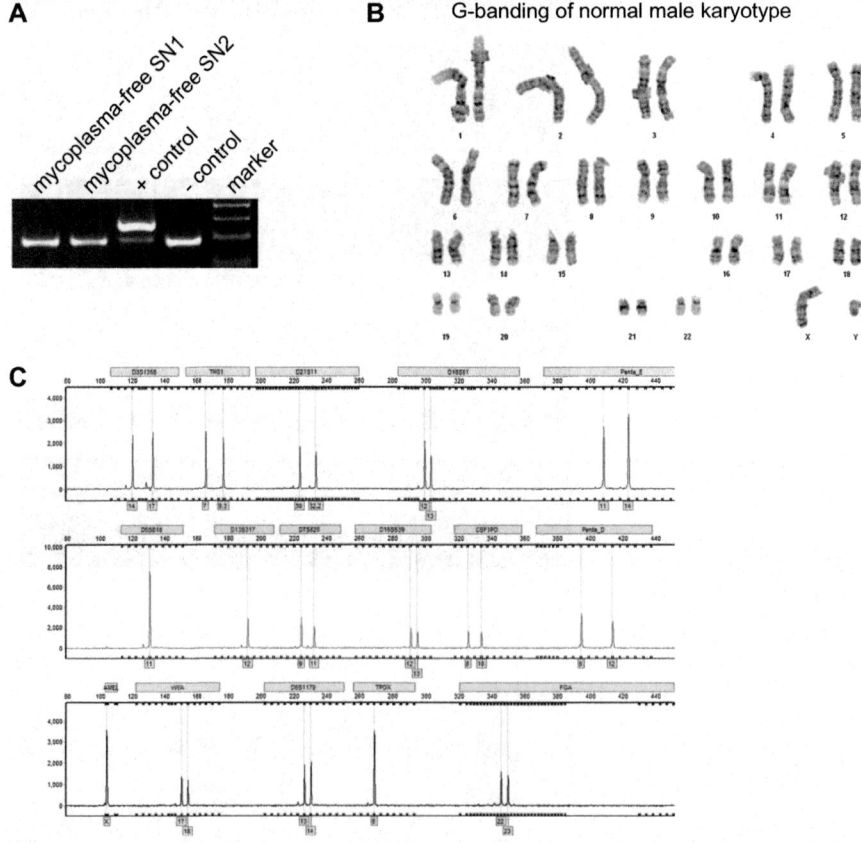

Fig. 2 Assay readouts for quality control. (**a**) Mycoplasma-specific PCR; (**b**) karyotyping using G-banding for genetic stability; (**c**) authentication of human cell lines is based on short tandem repeat (STR)

2.2 Pluripotency Testing by Immunofluorescence and Flow Cytometry

Key to the scientific and therapeutic potential of human PSCs is their capacity to differentiate toward cells of all three germ layers (endo-, ecto-, and mesoderm). Historically, pluripotency has been confirmed by conducting teratoma formation assays in severe combined immunodeficiency (SCID) mice and by using immunohistology to confirm the ability of PSCs to form all three germ layers in vivo. Because of its complexity and limited predictably, we have moved away from teratoma formation and now test PSCs for their capacity to differentiate toward all three germ layers by forming embryoid bodies (EBs). We use quantitative real time (qRT)-PCR to measure expression of a pluripotent marker gene signature predictive for pluripotency [17]. To test stem cells for their pluripotency, we also utilize phenotypic characterization methods such as immunocytochemistry (Fig. 1b) or flow cytometry (Fig. 1c). By using fluorescently labeled antibodies, the cultivated stem cells are stained for OCT4, NANOG, SSEA3, SSEA4, TRA-1-60, and TRA-1-81. These markers are expressed on PSCs and are commonly used to confirm pluripotency. Expression of these marker genes in iPSCs is an important quality parameter in achieving robust and stable cultivation conditions for PSCs. Furthermore, these surface markers are a useful indicator for predicting the ability of PSCs to differentiate toward specific cell lineages. The cell surface marker SSEA1 is an early differentiation marker that is not expressed at the pluripotent state of iPSCs [18, 19]. As a benchmark to confirm pluripotency, we accept a content of up to 5–8% of SSEA1-positive cells in our assays (Fig. 1c).

2.3 Genomic Stability of Human Pluripotent Stem Cells

Human PSCs can be maintained for several passages without developing any chromosomal abnormalities. However, with continued passaging, selection pressure can cause chromosomal abnormalities during cell growth.

To detect chromosomal abnormalities in PSCs, several methods are being used: (1) the traditional cytogenetic method, which analyzes G-banding of 20–30 metaphases (Fig. 2b), (2) fluorescent in situ hybridization (FISH), (3) spectral karyotyping (SKY), (4) single nucleotide polymorphism (SNP), and (5) copy number polymorphism mapping [20]. These methods differ in their resolution for detecting chromosome abnormalities. G-banded karyotyping can detect microscopic genomic abnormalities (5–10 Mb) such as inversions, duplications/deletions, balanced and unbalanced translocations, aneuploidies, and >10% mosaicism. It does not detect submicroscopic genomic abnormalities (<5 Mb).

SNP can detect genomic gains and losses (copy number variants), duplications/deletions, unbalanced translocations aneuploidies, loss of heterozygosity/absence of heterozygosity, and <20% mosaicism. It does not detect balanced translocations, inversions, >20% culture mosaicism, and chromosomal position of genomic gains.

Table 1 Karyotypes from 220 human PSC lines

	Number of samples for karyotype analysis	Percentage
Total	220	100
Normal	159	72.3
Abnormal	61	27.7

Table 2 Genetic abnormalities found in human PSC lines

	Type of abnormality ($n = 61$)	Percentage
Isochromosome 20q	43	70.5
Trisomy 12	12	19.7
Trisomy 8	4	6.5
Other	2	3.3

Within our quality control workflow and especially during cell banking, PSCs are regularly tested for chromosome abnormalities. Over the years, we have seen genomic abnormalities of up to 28% (Table 1) [21, 22]. These abnormalities could probably be explained by the different reprogramming methods that have been used and by the need to expand primary iPSC clones further. Analysis of the chromosomal abnormalities identified isochromosome 20q as the most common abnormality in our cultures [23, 24] (Table 2). Most probably, this mutation provides cells with a proliferative advantage and/or resistance to apoptosis. To avoid the development of chromosome abnormalities, PSCs should be cultivated for a limited cell passage number because there is a progressive tendency to acquire genetic changes during prolonged culture [25]. A report from the International Stem Cell Initiative showed that late-passage cultures of paired hPSC lines were approximately twice as likely to have a chromosome abnormality as those of early passage cultures [26]. There have also been reports of differences in the genetic stability of PSC lines. A recent study from a stem cell consortium with 711 iPSC lines showed copy number alterations in 41% of the lines [27]. Another study showed that 34% of 125 IPSC lines were karyotypically abnormal [26]. In agreement with the mentioned studies, our work has also shown high levels of genetic abnormalities (28%, Table 1). This suggest that before choosing a PSC line for further experiments one should check whether the line has been previously reported to show genomic instability.

2.4 Identity and Authenticity of Cell Lines Using Short Tandem Repeat Analysis

In today's dynamic scientific environment, cell lines are transferred between laboratories around the world. To ensure cell line authenticity and to prevent

Table 3 Short tandem repeat polymorphisms

Locus	Chromosomal location	ATCC marker	Alleles
D3S1358	Chr03	No	15/17
TH01	Chr11	Yes	9.3
D21S11	Chr21	No	31.2/32.2
D18S51	Chr18	No	12/15
Penta_E	Chr15	No	9/14
D5S818	Chr05	Yes	12/13
D13S317	Chr13	Yes	11/12
D7S820	Chr07	Yes	8/10
D16S539	Chr16	Yes	12/13
CSF1PO	Chr05	Yes	10/13
Penta_D	Chr21	No	9/13
AMEL	X/Y	Yes	X/Y
vWA	Chr12	Yes	16/18
D8S1179	Chr08	No	13
TPOX	Chr2	Yes	8/9
FGA	Chr04	No	24/26

accidental cross-contamination or mislabeling of cultures, they should undergo a regular identity test. For this reason, it is essential to validate a cell line using short tandem repeat (STR) analysis to confirm its identity. This is especially important for patient-specific iPSC lines to ensure that a specific cell line indeed corresponds with the disease-causing genetic mutation of the patient donor.

Genetic profiling of cell lines using multiplex PCR DNA STR profiling (Fig. 2c and Table 3) was performed for every PSC cell line in our laboratory on a regular basis. The STR analysis has a high degree of specificity. The preferred cell culture practice should be that only one cell line is processed at a time during handling in the biosafety cell culture cabinet. To protect the identity of the donor, data from the STR analysis should be handled confidential for ethical reasons.

3 Using Human iPSC-Derived Neural Cells as Drug Discovery Platforms

A decade after the development of human iPSC technology [9], iPSC-derived models have been implemented by the pharmaceutical industry for drug discovery programs. In particular, research on neurodevelopmental and neurodegenerative disorders has benefited considerably from iPSC-based disease modeling, especially because neurologically relevant cell types such as neurons used to be available only from post-mortem samples. Patient-specific iPSC-derived neural cells can resemble disease-relevant cellular phenotypes and be used as models for studying molecular mechanisms. Therefore, they possess great potential as drug screening platforms for identifying novel therapeutic targets (primary assay) or validating already existing preclinical drug candidates (secondary assay). The latest developments in iPSC-based screening platforms could provide clinical relevance for cell-based assays [28] and potentially reverse the current trend of decline in pharmaceutical R&D efficiency [3].

Indeed, iPSC technology platforms have already resulted in the first clinical candidates for treatment of neurological disorders. An example is a Tau-specific antibody from Bristol-Myers Squibb, currently being investigated in a Phase I safety trial [29]. Increased secretion of N-terminal fragments of Tau (eTau) into the extracellular supernatant from Alzheimer's disease patient iPSC neurons was observed compared with healthy control neurons, and led to development of the eTau-specific antibody [30].

Here, we describe the workflow in our laboratory for an established iPSC-derived drug screening platform for neurological disorders. To ensure that this strategy conforms to drug discovery protocols, we established a multistep workflow consisting of the following steps:

1. Whole human blood samples or skin biopsies were obtained from patients with neurological conditions.
2. Patient-specific iPSC lines were generated by reprogramming erythroblast or skin fibroblasts using the CytoTune-iPS Sendai Reprogramming Kit.
3. Derived iPSC lines were maintained in feeder-free conditions.
4. Neural progenitor cells (NPCs) were differentiated from hiPSCs using a dual SMAD inhibition protocol according to the literature [31] and implementing distinct modifications [32, 33] (Fig. 3a). The differentiation processes for generating neuronal cultures from hPSCs are depicted in Fig. 3b.

A crucial challenge of working with hiPSCs is their intrinsic propensity to spontaneous, undirected differentiation. To ensure a standardized and robust differentiation toward the targeted cell type, a stringent quality control workflow must be applied. In our laboratory, we rigorously test STR polymorphisms in patients' tissue samples, iPSC lines, and NPC lines to confirm cell line identities (Fig. 2c). Our iPSC lines and NPC lines are banked in sufficient quantities to enable their single use and to reduce the risk of genomic alterations that can be acquired by repetitive cell passaging. Our ability to bank NPCs as intermediate progenitors is crucial in reducing well-to-well variability in an iPSC-based drug discovery platform. Furthermore, the generation of neural progenitor cell banks is a prerequisite for scale-up of cell production toward cell numbers required for high-throughput screening campaigns. To run a phenotypic-based screening campaign with a library size of 1.1 million chemical compounds, a consistent cell supply over the period of 20 weeks was pivotal. We accumulated a total of 400 T175 cell culture flasks, which were required for providing 30 billion NPCs for use in the image-based assay. For this particular screening campaign, we derived neural progenitors and expanded those to generate a homogenous cell bank. This cell bank comprised more than 130 aliquots of 5 million NPCs per vial. The high-content imaging for

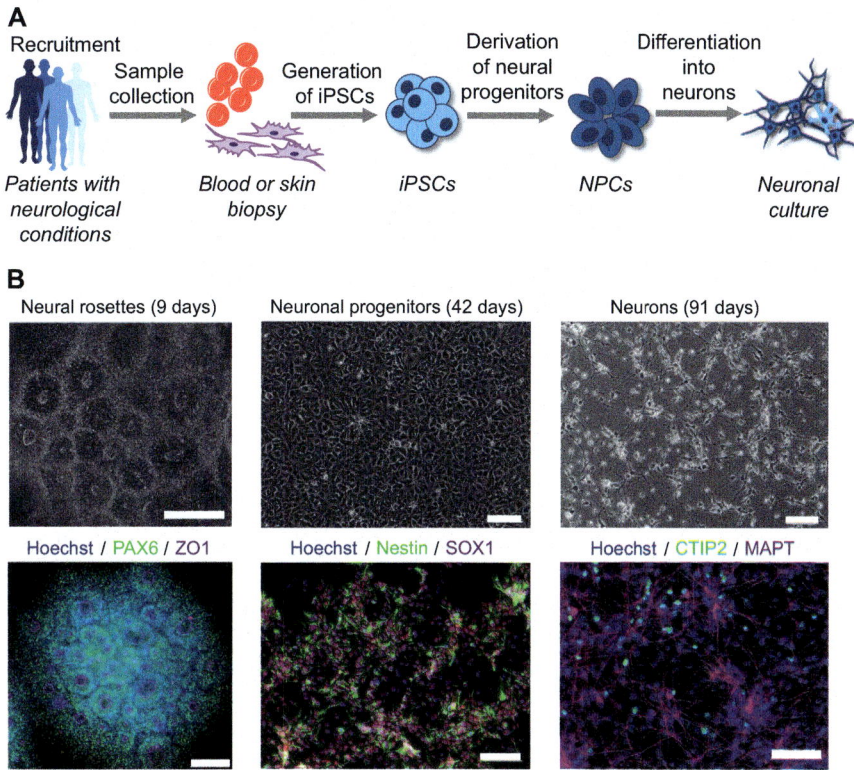

Fig. 3 iPSC-derived neuronal-cell-based drug discovery platform. (**a**) Flowchart depicting iPSC-derived neuronal differentiation. (**b**) Microscopic images of different stages during neuronal differentiation

quantification of neurite outgrowths served as primary readout for this type of in vitro human neurogenesis screen. High-content screening was performed in a 384-multiwell plate format. This image-based assay to identify chemical compounds that are capable of stimulating neurogenesis turned out to be highly sensitive to subtle differences in cell density. Over the course of neuronal differentiation from PSCs, small differences in the initial cell seeding number can accumulate and result in pronounced well-to-well variability. As a consequence, we generated NPC lines. The use of these pre-committed neural progenitor cells as starting point for our screening efforts shortened the assay time significantly. Specifically, it was possible to omit the initial differentiation step from hPSCs, namely neural induction and the neural epithelial enrichment phase, leading to significant shortening of the cultivation time by 6 weeks and reduced well-to-well variability.

The genomic integrity and cell type identity of iPSC lines and NPC lines can be verified by their characteristic cell morphology and cell type specific marker expression pattern, before being adapted to drug discovery platforms. This characterization is part of our standard assay quality assurance efforts. During the last step of our workflow, predifferentiated NPCs are plated onto screening-compatible 384-well plate formats. Subsequently, the NPCs undergo a neuronal differentiation period of 35–49 days. Depending on the biological question, neural cell cultures can be used for compound screenings or for cellular and molecular phenotyping. Furthermore, cell-based assays for phenotypic- or target-based screening can employ multiple assay readouts such as imaging-based and gene or protein expression-based analyses. To measure functional neuronal activity, we applied calcium imaging, patch-clamping, and multi-electrode arrays as readouts [32].

4   Conclusion and Perspectives

Over the last few years, the use of PSC-derived cell assays has emerged as a very attractive tool for the pharmaceutical industry. In this chapter, we have described a workflow for successful use of PSCs in drug discovery. Our data underlines the importance of stringent quality control and use of defined cultivation and differentiation methods in an established workflow. A defined process during every stage is needed to ensure the highest quality and reproducibility for drug discovery assays. This process includes patient recruitment, sample collection, generation of iPSCs and their differentiation into specific cell lineages, assay development, and screening. At multiple stages during this process, quality requirements such as STR profiling, pluripotent marker validation, genomic integrity, and cell banking should be implemented within a standardized workflow. To ensure that protocols and assay formats generate reproducible results across laboratories, the initiation of multisite experiments within large research consortia can help to establish best practices throughout the scientific community. StemBANCC (http://stembancc.org/) is a large-scale academia–industry partnership in the area of stem cell research. It brings together 35 partners that integrate a consortium with pharmaceutical companies, research institutions, and small and medium enterprises to exploit the expertise across sectors and enhance knowledge transfer between academia and industry. Participants within the consortium establish protocols for patient sample handling, reprogramming, lineage specific differentiation, and assay development in a consistent and transparent way. The combination of such a defined workflow with stringent quality management, together with efficient methods to generate defined cell types [34], will provide the basis for success in using PSC-derived disease-relevant cellular models for drug discovery.

Acknowledgements The authors acknowledge Silke Zimmermann, Nadine Dahm, and Corinne Marfing for technical assistance and Cecilia Cariño Morales for proof reading.

References

1. Swinney DC, Anthony J (2011) How were new medicines discovered? Nat Rev Drug Discov 10:507–519. https://doi.org/10.1038/nrd3480
2. Rees S, Gribbon P, Birmingham K, Janzen WP, Pairaudeau G (2016) Towards a hit for every target. Nat Rev Drug Discov 15:1–2. https://doi.org/10.1038/nrd.2015.19
3. Horvath P et al. (2016) Screening out irrelevant cell-based models of disease. Nat Rev Drug Discov 15:751–769. https://doi.org/10.1038/nrd.2016.175
4. Thomson JA et al. (1998) Embryonic stem cell lines derived from human blastocysts. Science 282:1145–1147
5. Ciampi O et al. (2016) Generation of functional podocytes from human induced pluripotent stem cells. Stem Cell Res 17:130–139. https://doi.org/10.1016/j.scr.2016.06.001
6. Patsch C et al. (2015) Generation of vascular endothelial and smooth muscle cells from human pluripotent stem cells. Nat Cell Biol 17:994–1003. https://doi.org/10.1038/ncb3205
7. Tran TH et al. (2009) Wnt3a-induced mesoderm formation and cardiomyogenesis in human embryonic stem cells. Stem Cells 27:1869–1878. https://doi.org/10.1002/stem.95
8. Takahashi K, Yamanaka S (2006) Induction of pluripotent stem cells from mouse embryonic and adult fibroblast cultures by defined factors. Cell 126:663–676. https://doi.org/10.1016/j.cell.2006.07.024
9. Takahashi K et al. (2007) Induction of pluripotent stem cells from adult human fibroblasts by defined factors. Cell 131:861–872. https://doi.org/10.1016/j.cell.2007.11.019
10. Musunuru K (2013) Genome editing of human pluripotent stem cells to generate human cellular disease models. Dis Model Mech 6:896–904. https://doi.org/10.1242/dmm.012054
11. Ding Q et al. (2013) A TALEN genome-editing system for generating human stem cell-based disease models. Cell Stem Cell 12:238–251. https://doi.org/10.1016/j.stem.2012.11.011
12. Ding Q et al. (2013) Enhanced efficiency of human pluripotent stem cell genome editing through replacing TALENs with CRISPRs. Cell Stem Cell 12:393–394. https://doi.org/10.1016/j.stem.2013.03.006
13. Terstegge S et al. (2009) Laser-assisted selection and passaging of human pluripotent stem cell colonies. J Biotechnol 143:224–230. https://doi.org/10.1016/j.jbiotec.2009.07.002
14. Ludwig TE et al. (2006) Derivation of human embryonic stem cells in defined conditions. Nat Biotechnol 24:185–187. https://doi.org/10.1038/nbt1177
15. Mariotti E, Mirabelli P, Di Noto R, Fortunato G, Salvatore F (2008) Rapid detection of mycoplasma in continuous cell lines using a selective biochemical test. Leuk Res 32:323–326. https://doi.org/10.1016/j.leukres.2007.04.010
16. Young L, Sung J, Stacey G, Masters JR (2010) Detection of mycoplasma in cell cultures. Nat Protoc 5:929–934. https://doi.org/10.1038/nprot.2010.43
17. Muller FJ, Brandl B, Loring JF (2008) StemBook [Internet]. Harvard Stem Cell Institute, Cambridge. doi:https://doi.org/10.3824/stembook.1.84.1
18. Adewumi O et al. (2007) Characterization of human embryonic stem cell lines by the international stem cell initiative. Nat Biotechnol 25:803–816. https://doi.org/10.1038/nbt1318
19. Wiese C, Kania G, Rolletschek A, Blyszczuk P, Wobus AM (2006) Pluripotency: capacity for in vitro differentiation of undifferentiated embryonic stem cells. Methods Mol Biol 325:181–205. https://doi.org/10.1385/1-59745-005-7:181
20. Martins-Taylor K et al. (2011) Recurrent copy number variations in human induced pluripotent stem cells. Nat Biotechnol 29:488–491. https://doi.org/10.1038/nbt.1890
21. Baker D et al. (2016) Detecting genetic mosaicism in cultures of human pluripotent stem cells. Stem Cell Rep 7:998–1012. https://doi.org/10.1016/j.stemcr.2016.10.003
22. Taapken SM et al. (2011) Karotypic abnormalities in human induced pluripotent stem cells and embryonic stem cells. Nat Biotechnol 29:313–314. https://doi.org/10.1038/nbt.1835
23. Avery S et al. (2013) BCL-XL mediates the strong selective advantage of a 20q11.21 amplification commonly found in human embryonic stem cell cultures. Stem Cell Rep 1:379–386. https://doi.org/10.1016/j.stemcr.2013.10.005

24. Nguyen HT et al. (2014) Gain of 20q11.21 in human embryonic stem cells improves cell survival by increased expression of Bcl-xL. Mol Hum Reprod 20:168–177. https://doi.org/10.1093/molehr/gat077
25. Liu P et al. (2014) Passage number is a major contributor to genomic structural variations in mouse iPSCs. Stem Cells 32:2657–2667. https://doi.org/10.1002/stem.1779
26. Amps K et al. (2011) Screening ethnically diverse human embryonic stem cells identifies a chromosome 20 minimal amplicon conferring growth advantage. Nat Biotechnol 29:1132–1144. https://doi.org/10.1038/nbt.2051
27. Kilpinen H et al. (2017) Common genetic variation drives molecular heterogeneity in human iPSCs. Nature 546:370–375. https://doi.org/10.1038/nature22403
28. Scannell JW, Blanckley A, Boldon H, Warrington B (2012) Diagnosing the decline in pharmaceutical R&D efficiency. Nat Rev Drug Discov 11:191–200. https://doi.org/10.1038/nrd3681
29. Mullard A (2015) Stem-cell discovery platforms yield first clinical candidates. Nat Rev Drug Discov 14:589–591. https://doi.org/10.1038/nrd4708
30. Bright J et al. (2015) Human secreted tau increases amyloid-beta production. Neurobiol Aging 36:693–709. https://doi.org/10.1016/j.neurobiolaging.2014.09.007
31. Chambers SM et al. (2009) Highly efficient neural conversion of human ES and iPS cells by dual inhibition of SMAD signaling. Nat Biotechnol 27:275–280. https://doi.org/10.1038/nbt.1529
32. Costa V et al. (2016) mTORC1 inhibition corrects neurodevelopmental and synaptic alterations in a human stem cell model of tuberous sclerosis. Cell Rep 15:86–95. https://doi.org/10.1016/j.celrep.2016.02.090
33. Dunkley T et al. (2015) Characterization of a human pluripotent stem cell-derived model of neuronal development using multiplexed targeted proteomics. Proteomics Clin Appl 9:684–694. https://doi.org/10.1002/prca.201400150
34. Qi Y et al. (2017) Combined small-molecule inhibition accelerates the derivation of functional cortical neurons from human pluripotent stem cells. Nat Biotechnol 35:154–163. https://doi.org/10.1038/nbt.3777

Index

Copy number variant (CNV), 190, 195, 196, 199
CRISPR-associated (Cas) systems, 172–174
  research applications, 177–178
Cultivation methods, human PSCs, 210–215
Culturing method, 198
Cytogenetic method, 195, 213

D

Decellularized matrices, 160, 161, 163, 164
Diabetic cardiomyopathy (DCM), 57
Disease modeling and drug development, 179
Disease-related cellular models, 208
DNA damage, 192
DNA double-strand breaks (DSBs), 172, 174, 175, 179
DNA repair mechanisms
  homologous recombination, 172
  nonhomologous end joining, 172
DNA replication
  DNA virus-derived sequences, 7
  episomal plasmids, 6–7
  inhibitors, 198
DNA sequence comparisons, parental somatic cells and hiPSCs, 194
DNA transfection techniques, 171
DNA virus-derived sequences, 7
Droplet Digital PCR, 11
Drug discovery
  biological pathways, 208
  biological target identification, 208
  cellular assays, 208
  high-throughput screening, 208, 209
  human iPSC-derived neural cells, 215–218
  pharmaceutical industry, 208, 210
  protocols, 216
  PSC-derived disease-relevant cellular models, 218
  quality control requirements, 210–215
Drug screening assays, 56

E

EBV origin of replication (oriP), 7
Ectodermal lineage
  conversion strategies, 80–82
  neuronal cells, 80–82
Ectopic MyoD expression, 5
Electroporation protocols, 174
Embryoid bodies (EBs), 52, 213
  endothelial differentiation methods, 152
  vSMC differentiation, 152–153
Embryonal carcinoma cells (ECCs), 3

Embryonic tissue transplantation studies, 3
Endodermal lineage
  conversion strategies, 83–84
  hepatocytes, 82
  insulin-positive β-cells. 84, 89
Endothelial cells (ECs)
  coagulation, 150
  EB formation, 152
  immuno/inflammatory response, 150
  network formation, 156
  pathogen, barrier, 150
  physiological processes, 150
  sprouting and migration, 156
  and vSMC differentiation
    embryoid body formation, 152–153
    OP9 Co-culture, 152
    2D monolayer differentiation, 153
  wound healing, 150
Engineered cardiac tissue function, 131–133
Enhanced programming efficiency, 90
Enhanced specificity SpCas9 variants, 174
Enzymatic/non-enzymatic bulk passaging methods, 197
Epigenetic change, 192–193
Episomal reprogramming systems, 7–11, 15, 193–194
Episomal vectors, 201
Epithelial-to-mesenchymal (EMT) transition, 42
Epstein Barr virus (EBV)-derived episomal plasmids, 7
Epstein Barr virus nuclear antigen 1 (EBNA1) protein, 7
Ex vivo gene therapy, 176, 182

F

Familial cardiac hypertrophy (FCH), 57
Fatty acid β-oxidation, 58
FCH patient-derived hiPSC-CMs, 57
Fetal calf serum (FCS), 52, 53
Fluorescent in situ hybridization (FISH) analysis, 190, 195, 197, 199, 213
Fluorescent protein expression, 180
Functional blood vessels
  arteries, 150
  arterioles, 150
  capillaries, 150
  endothelial cells, 150
  pericytes, 150
  vascular smooth muscle cells, 150
  veins, 150
  venules, 150
Functional neuronal activity, 218